Biomaterials

Series Editor
Marie-Christine Ho Ba Tho

Biomaterials

Edited by

Véronique Migonney

iSTE

WILEY

First published 2014 in Great Britain and the United States by ISTE Ltd and John Wiley & Sons, Inc.

ISTE Ltd
27-37 St George's Road
London SW19 4EU
UK

www.iste.co.uk

John Wiley & Sons, Inc.
111 River Street
Hoboken, NJ 07030
USA

www.wiley.com

Library of Congress Control Number: 2014947880

British Library Cataloguing-in-Publication Data
A CIP record for this book is available from the British Library
ISBN 978-1-84821-585-6

Contents

Introduction

The aim of writing this book is to give a general and clarified overview on the biomaterials field. This book pretends to allow people to respond to some questions such as: what are biomaterials? Why are they used? When is it necessary to use them?

In addition to these very simple questions, but seldom well understood, the restrictions due to the biological environment and the implantation in the body also have to be well known. What does "biocompatibility" mean? What are the restrictions of the use of biomaterials? These questions and corresponding response are decisive for the appropriate and safe use of biomaterials as implants and prostheses. Directives, rules and certification processes have to be taken into account to certify the reliability of biomaterial implants and to prevent further problems from occurring, such as the recent highlighted problem of the silicone breast implants, or those of the cytotoxicity of some debris generated in the case of particular joint implants etc. Other problems exist and are even less known; they have to be seriously taken into account.

To conclude, this book is proposed to help simplify the definition and the overview of the biomaterials fields and does not claim to compete with major detailed books in the domain of biomaterials; these have to remain the references and were written and edited by

---

Introduction written by Véronique MIGONNEY.

certain founding fathers of biomaterials science: *Biomaterials Science: An Introduction to Materials in Medicine* by B. Ratner, A. Hoffman, F. Schoen and J. Lemons and the *Dictionary of Biomaterials* by D. Williams.

This short book is simply presented in order to be accessible to anyone interested in biomaterials. It arose from the numerous questions I received not only from students but also from colleagues, scientific or not, foreign or not to the domain of biomaterials, which showed that the biomaterials field is not well understood. It attempts to prevent some classical confusion regarding biomaterials and its field of applications. It is voluntarily short and will refer to more important work when it will be necessary.

I truly thank my colleagues who accepted to contribute to this work, my "fathers" in biomaterials science, the editor who trust on me inviting me to write this book and my family who is every day living with a researcher and teacher obsessed by biomaterials and all its correlated interrogations.

1

History of Biomaterials

1.1. Introduction

The aim of this chapter is to give an outline of the long, progressive and amazing history of biomaterials.

To begin the chapter entitled "History of biomaterials" it is necessary to give a first simple definition of the term "biomaterials":

"Biomaterials are materials intending to supply all or part of a deficient organ".

This quite restricted definition – which will be improved throughout this book – gives an idea and an overview of the goals, needs and even potential applications of biomaterials. We will discover that even through the use of biomaterials dates from a very long time ago, the notion of biomaterials science emerged in the 1960s due to the agreement of a few of open-minded scientists wishing to build this new domain of science. Then, the word "biomaterials" started to be used at the same time as the birth of learned societies in this domain: European and American researchers worked a lot to build the American and European Societies for Biomaterials. This will be developed later.

Chapter written by Véronique MIGONNEY.

1.2. The evolution of biomaterials: several generations

Today, we can briefly differentiate four generations of biomaterials:

1) The first generation started with humanity and was simply restricted to the materials which were available in the natural environment of the human being and used to simply repair organs.

2) The second is very long and started with the "history" of the improvement of the human knowledge and finished in the middle of the 20th century – this generation gained all the benefits from the industrial revolution.

3) The third and actual generation started with the "birth" of polymers as new and promising materials added to the fruit of scientists' knowledge and research about materials and the way they can be transformed or elaborated and the extraordinary living systems.

4) The fourth generation is in its beginnings and is a mixture of dreams and realities.

Even if these four generations of biomaterials have been identified, the duration of each one is very variable. The last 20 years have led to so much progress in the disciplines involved in biomaterials science that it is very difficult to schedule the duration of the last generation, as well as imagining what the next one will be.

The use of biomaterials for therapeutic purposes such as the repairing of wounded organs began a very long time ago. It was at first strictly linked to the accessibility of people to "materials", their knowledge of the materials' properties as well as the methods and processes they used to transform and develop them. As it is difficult to date the beginning of the uses of biomaterials, this chapter will intend to describe what we know about the first materials used with the aim of repairing an organs or a part of an organ. We will also discover the unlimited imagination of human beings as well as the diversity of materials they used as "biomaterials" to prevent a deficiency and/or to repair unhealthy or wounded tissues or organs.

To date, the use of biomaterials seems to have started from antiquity and probably since the origins of human beings. The evidence of their use as implants or prostheses were mainly discovered during the two last centuries on human skeletons or skulls during the excavations of sites which were attributed to different civilizations of antiquity: Egyptian, Roman, Greek and Etruscan. It is worthy to note that the literature is prolix about the "probable" uses of biomaterials for dental applications by the antique civilizations and even the following ones. This is probably due to the fact that since antiquity, the proof of its use for dental applications were the most numerous and/or variously encountered: wire to link teeth together, teeth implants, filling materials, etc. In contrast, despite the full of imagery vision of the replacement of injured or amputated limbs by wooden pieces, it is quite difficult to find very ancient examples of other medical applications and real uses of biomaterials such as in ophthalmology, vascular surgery, orthopaedic surgery, etc. Such examples would probably have taken place a long time after antiquity. The question which arises is: did these materials really exist as biomaterials – therapeutic purposes to replace missing teeth or limbs of living men or were they only used to improve the esthetical aspect of remains and of graves? The response to the last question is of importance to certify the exact beginning of the use of biomaterials for organ repairing.

Only historians can date the beginning of the existence of "biomaterials" and bring the proof of its literal use, that is material to mitigate a deficiency of all or part of a hurt or wounded organ. As a matter of fact, recently questioned historians said that they had never heard about the use of biomaterials in antiquity. Despite this interrogation something in the history of biomaterials sounds very strange. Whereas in the first civilizations human beings had yet to imagine their uses and found the necessity of their use in aiding deficiencies of some of their organs or parts of organs, today the term "biomaterials" is neither correctly or currently used except for the "specialists of the domain" although everybody in his/her life will have to face its use and to understand how it works. Therefore, the existence of biomaterials is very ancient while the science of biomaterials appeared very recently.

From the outset, human beings resorted to biomaterials to improve their conditions of health. It is very difficult to make a clear and precise classification of the uses of biomaterials because much is missing and it is clear that most of the work still has to be achieved. The aid of the historian community would be helpful to enlist, to propose and to validate a classification for: a time schedule; application of the device; choice of the material. A proposed "easy classification" is made here, looking for different materials, which were used by human beings in various applications; then this classification will mix applications and materials. The aim is to show that applied research in the biomaterials field started with the existence of men and was the fruit of their imagination and of their intelligence.

1.3. Was gold the first "biomaterial"?

Was gold the first "biomaterial"? The response to the question is not so simple. The use of materials for dental repair could have started in antiquity when men used diverse natural organic and/or inorganic materials such as bones, animal teeth or even wood to replace missing teeth.

Based on the literature, one of the "first biomaterials" could have been a hybrid material made of an animal tooth linked to the patient's teeth by a gold wire. This "first" dental prosthesis had been attributed to the Etruscan civilization and could be dated around 2,600 years ago. It was used to replace the upper incisors with a cow tooth fixed to the neighboring teeth by a gold wire .

The use of gold is not surprising since, gold being a noble metal, it exhibits a relative chemical inertness and an excellent resistance to chemical attacks because it is not sensitive to oxidation. This explains its prolonged use in dental applications and the fact that this metal is still considered as a "gold standard" in dentistry. The "chemical" properties added to its ductility and malleability easily explained the importance of its use in many medical applications.

The way gold material was used for dental repairing seemed to be very different between the occidental and the oriental societies

[SCH 98]. The differences reported by Bardinet in his history of dental prosthesis [BAR 90] were shown on skeletons and found to be technical in nature. Then, despite these different techniques carried out, Bardinet concluded and assumed that the unique and common aim of using gold was to consolidate teeth anchorage: gold is one of the first "biomaterials" used in former civilizations.

Due to these reported observations, studies and conclusions, until the 18th century, dental prostheses consisted of human teeth (extracted from patients) bound by golden wire ligatures. It is necessary to wait to the end of the 19th century for the appearance of gold teeth prostheses elaborated by imprint and casting techniques [VAN 85]. The dentist Aguilhon de Sarran from the "French Society of Stomatology" proposed this application of gold in 1903: having taken a tooth imprint, he melted gold and poured it into the mould to obtain an inlay in gold. Some years later, Solbrig developed this technique; today the gold inlay remains the oldest effective closing technique. This active and fruitful period in terms of innovation in dental repair was followed by the development of the dental crown technique; for its many advantages and lack of weaknesses, gold-based alloys were commonly used. Today, and mostly for aesthetic reasons, the metallic parts (gold or other metallic materials) are masked by using ceramic materials since their esthetical appearance imitates natural teeth well .

Therefore, to conclude on gold, this metal appears to be one of the first and principal materials used in by ancient civilizations and, incredibly, it is still used today.

Despite gold being used in dental repair, the examples are numerous and varied on the use of other materials than gold for dental applications:

Roman civilization: in Roman times, the dental filling materials consisted of crushed slate, and lead combined, or not, with wool or with gold. Missing teeth were still replaced by ivory or bones maintained by gold wire sheets linked to the other surrounding teeth.

Arabian civilization: following discoveries dating from antiquity, the middle ages gave more detailed information on the potential use of biomaterials. Rhazes (865–932 AD Persia), considered to be the first Arabian medical doctor, was a surgeon and a chemist and paid particular attention to tooth repair. He described some hygiene requirements and even set up a formula of a mastic based on resin to fill dental cavities [RHA 80]. Despite this activity related to biomaterials, this surgeon was most known for his huge encyclopaedia entitled "Continens".

In the 10th century, other Arabian dentists as Abou Amed Gaafar, Halid al-Gazzar and Avicenne described "the first treatments" of tooth decay using different materials to fill up the dental cavity. Those materials: oak gall, dyes, resin of pine and cedar, cypress, myrrh, honey, medicinal herbs, pepper, camphor, drugs for the pain, arsenic and milk of she-wolves can be considered as biomaterials [COR 89].

Therefore all these reported studies are clues of the merit that goes back to Arabian civilization for the progress of medicine and also for the use of materials as "biomaterials". Translating Greek scientific texts, Arabians allowed the occidental civilization to know and learn a lot all about ancient observations [BAR 90, BEL 20, BEC 99, CLA 34.

European civilization: Fauchard, considered in France as the father of modern dentistry, in his treaty of anatomy published in 1728 gives different recommendations on tooth repair and describes a (bio)material for tooth filling made of tin, lead and gold. At the same time, the prostheses were set up using gold but also enamelled metals and porcelain, which started to be used.

1.4. The use of glass to replace eyes

As for the other applications of biomaterials, throughout the years and from any given time, human beings were looking for materials able to replace "missing" or "wounded" eyes. In Egypt, the excavations of funeral sites showed that precious stones or painted glass were used to replace missing eyes. As mentioned above for gold wire use in dental repair application, no proof can be given of the time those precious stones and glass were used: during peoples' lives or after their death?

The first "functional" eye prostheses – used during the a person's life – appeared at the end of the 16th or at the beginning of the 17th Century. Eye prostheses were made of gold or silver balls decorated with a painted porcelain iris. Due to the density of gold and silver, those eye prostheses were very heavy and uncomfortable. They were replaced at the 18th Century by French prostheses made of half-cut shells, better taking the shape of the ocular cavity. However, the quality of those prostheses was very bad and their aesthetic appearance was poor. Around the middle of the 18th Century, German glass blowers found how to improve the quality of these prostheses but it was to the detriment of its "biocompatibility" because the glass contained "lead" which induced intense irritations to the patients and at the same time it degraded within months. As the prostheses had to be frequently replaced, glass blowers worked to set up a higher quality of glass to fabricate longer-term prostheses. Nowadays, this kind of prosthesis still exists and is used for patients affected by microphthalmia (developmental disorder of the eye) – microphthalmia literally means "small eye". Today eye prostheses are made of glass or of synthetic resins and are specially elaborated and fabricated for each patient.

1.5. Wood, leather, stainless steel to replace amputated limbs

One of the most well known orthopaedic prostheses would come from Egypt and date from antiquity. This prosthesis – visible at the museum of Cairo – was described by German researchers assuming that Egyptians were able to amputate limbs and design prosthesis. They hypothesized that through the observation of an old Egyptian mummy of a dead woman – approximately 3000 years ago – she was amputated at the right toe and would have had a big toe prosthesis made of sculptured wood and connected to the foot by a sewn leather girdle and fabrics. The tracks of wear debris would show that the prosthesis was used and allowed this woman to walk. Besides this example, other people dated the first prostheses from prehistory, when people began to walk on their two feet. To survive, they needed to find solutions how to replace amputated limbs or distorted limbs.

During antiquity, the Greeks and Romans also made prostheses. Evidence is more numerous, due to the Greek historians who wrote narratives, such as Herodotus who described one of his heroes being amputated at the leg and who therefore wore a wooden leg.

In the Middle Ages, prostheses such as pestles and hooks were made of wood and metals and had a functional purpose being used for fights. Nevertheless, their weight and lack of quality as well as properties and sophistication limited their used to only few people with the idealized image of knights and pirates.

In the period of the Renaissance, prostheses saw a new development at the same time of the revival of the sciences, of medicine and of surgery – Ambroise Paré worked on the development of arm and hand prostheses.

Nevertheless, it is during the two last world wars and because of the soldiers' numerous amputations after battles that the design of prostheses has been changed. Prostheses were realized in wood or metal with sockets made of leather reinforced by metallic pieces (probably stainless steel). These prostheses had numerous inconveniences such as their weight, the inflammation, irritations and allergies they induced at the level of the stub, and the deformability of the metal. Other materials such as aluminum and polymers were proposed – the first to relieve the patient from the weight of the prosthesis and the second to replace leather [FIS 00].

In 1950, due to their excellent and variable properties, polymers took a major place in the design and elaboration of prostheses. They were then made of polyester resins, polymers such as silicon and metal being more resistant, less cumbersome and almost unnoticed. The remaining problems were still the same: allergies, inflammations etc. because at that time the notion of biocompatibility and required properties for biomedical applications was not known.

Much progress is still being made and concerns, the design, the weight and resistance through the choice of the materials (Kevlar and/or composite for the socket), the type of prostheses such as “contact prostheses” for young sportive people and “sport prosthesis as running blades” such as those worn by the sadly notorious Oscar Pistorius, nicknamed “Blade Runner: the fastest man on no legs”.

1.6. Conclusions

In the brief history of biomaterials, it can be shown that since the origin or at least at the beginning of “history”, human beings have used various materials to replace organs or parts of organs. Numerous accounts have recorded the uses of materials for various replacements: glass for eyes, wood for teeth etc. Romans, Chinese and Aztecs used gold in dentistry; Egyptians and Indians used linen for suture by (also horsehair, cotton…). These materials were those to which they had access or which they used in everyday life. This was the case of the first generation of biomaterials.

Through this rapid historical overview of the “first generation of biomaterials” used with the aim of replacing wounded organs such as teeth, eyes and limbs, the most extraordinary conclusion is that some of the materials chosen by the ancients are still used today. In addition, this enables us to pay a tribute in the intelligence of people who, with the few means they had, contributed to the birth of biomaterials and to this fascinating domain of science, the purpose of which is to improve the life of the patients.

The further generations took less time to arise and their duration is shorter and they cannot be summarized in a short chapter. This is the reason this chapter is restricted to the simple use of a biomaterial and will not describe the most recent advances. Indeed, during the 20th century, the development of the plastics industry as well the increased interest in the ceramic domain opened the way to new materials, endowed with physicochemical and mechanical properties as diverse as they are interesting. “Polymers” or plastic materials, ceramic and metals will be detailed further in the next chapters.

1.7. Bibliography

[BAR 90] BARDINET T., "La prothèse dentaire au temps des pharaons mythe ou réalité?", *Inf. Dent.*, vol. 29, pp. 2553–55, 1990.

[BEC 99] BECKER M.J., "Etruscan gold dental appliances: three newly "discovered "Examples", *American Journal of Archaeology*, vol. 103, pp. 103–111, 1999.

[CLA 34] CLAWSON M.D., "Appareil de prothèse dentaire du Vème siècle avant J.C", *La Semaine Dentaire*, vol. 8, pp. 157–171, 1934.

[COR 89] CORRUCINI R.S., PACCIANI E., "Orthodontistry" and dental occlusion in Etruscans", *Angle Orthodontics*, vol. 59, no. 1, pp. 61–64, 1989.

[FIS 00] FISCHER L.P., PLANCHAMP W., FISCHER B., *et al.*, "Les premières prothèses articulaires de la hanche chez l'homme (1890–1960)", *Histoire Des Sciences Médicales*, vol. 34, no. 1, pp. 57–70, 2000.

[QUE 75] QUENOUILLE J.-J., "Les prothèses dentaires égyptiennes: Une légende qui a la vie dure", *Chir. Dent. Fr.*, vol. 269, pp. 45–46, 1975.

[RHA 80] RHAZE H., Continens 1280 Paris BNF.

[SCH 98] SCHNEIDER H., La prothèse dentaire dans l'Antiquité, thesis, 20 May 1998, Université Paris 5 René Descartes, Faculté de chirurgie dentaire.

[VAN 85] VAN MARTER J.G., *Some Evidences of Prehistoric Dentistry Independent Practitioner*, vol. 6, pp. 1–5, 1885.

[WEI 41] WEINBERGER B.W., "The Etruscan cast 'gold' crowns illustrated by Guerini", *Journal of American Dental Association*, vol. 28, pp. 1853–1854, 1941.

2

Definitions

2.1. Introduction

The previous chapter was devoted to the history of the very first generation of biomaterials, those to which human beings had access or which they used in everyday life. Since the beginning of the 20th Century, the development of scientific knowledge and the advances in areas such as materials and design methods has opened the path to the "second generation" of biomaterials exhibiting multiple and various valuable properties. Therefore, numerous metallic, ceramic and polymeric materials have shown themselves to be excellent candidates for numerous biomedical applications – ophthalmology, dental, cardiovascular or orthopaedic surgery. These materials allowed the realization of a large number of biomaterials used today as implants, prostheses and medical devices.

After having evidenced that human beings were very early confronted with the necessity of using materials for replacing injured or deficient limbs or organs, this chapter will show that for the use of these materials to be successful they have to answer numerous defined properties and requirements included under the term of "biocompatibility". Wood, glass, gold and various other materials were cleverly used as solutions for organ repairing at a time when nothing was known or described about the host response induced by these "non-controlled" (bio) materials. Even if human beings through

---

Chapter written by Véronique MIGONNEY.

this long period of time progressed in their knowledge and practice of material properties to find the right material for the right (specific) application, the notion of biocompatibility and control of the host response did not exist.

A large part of the terminology used in the biomaterials domain is not currently used in the everyday life and will be defined, detailed and/or explained. Different existing definitions will be proposed to understand the significance or to differentiate the words such as biomaterial, medical device, implant, prosthesis, tissue engineering and regenerative medicine. In addition, the notion of biocompatibility will be briefly introduced; one chapter will be devoted to biocompatibility through its different aspects: medical, legal and research.

2.2. Definitions of a "biomaterial"

What exactly does "biomaterial" mean? There are several senses which depend on who uses it: researcher in the domain, faculty, student, industrialist, medical professional, non-scientific or non-specialist population. Unfortunately, the meaning given to the word "biomaterial" may often be erroneous.

Since a word must have a meaning, the definitions are few: the definition found in the dictionary, the aim of which is to give a clear and correct idea of the sense of this word to non-specialist people. Based on the short life of biomaterials science, this word appeared only very recently in the dictionary, the definition given by the founding scientists of the science of biomaterials which is the most precise one, along with the advances and progresses in this new science, the definition has changed a little. Therefore several complementary definitions could be essential for understanding this recent and continuously evolving field.

2.2.1. *Dictionary definitions*

1) Larousse dictionary

Biomaterial: "substance or material intended to be implanted in a living body to replace an organ or a tissue. Prostheses, of the simplest (dental) in the most complicated (heart valve), are made with biomaterials".

2) Collins dictionary

Biomaterial: "any synthetic material used in prostheses or the replacement of natural body tissues".

2.2.2. *Definitions of biomaterials from expert scientists of the domain*

It is difficult to precisely date the first use of the word "biomaterial". It is probably during in the early 1970s following scientific meetings such as the Clemson University Biomaterials Symposia in the United States and the Conference of Chester in Europe. These conferences respectively led to the creation of the Society for Biomaterials (SFB) in 1975 and the European Society for Biomaterials (ESB) in 1976. These societies are members of the International Union of Societies for Biomaterials Sciences and Engineering (IUS-BSE) created in 1979 and showing that biomaterials science rapidly acquired a worldwide interest.

1) 1st definitions

Europe – The consensus after the conference of Chester led to the definition given by the Pr D. Williams. He defined a biomaterial as the following: "a nonviable material used in a medical device, intended to interact with biological systems".

United States – The Clemson University Advisory Board for Biomaterials has defined a biomaterial to be: "a systematically and pharmacologically inert substance designed for implantation within or incorporation with living systems".

2) Intermediate definition: "Any material, natural or not, including all or part of a living structure or a biomedical device which executes or replaces a bodily function."

These definitions of a biomaterial did not include only the artificial or synthetic biomaterials as metals, ceramics and polymers. A biomaterial can also be any graft (autograft, allograft and even xenograft) used as a transplanted material.

2.2.3. *Up-to-date definition of biomaterials*

The up-to-date definition of a biomaterial is the following: "*material intended to supply or to replace all or a part of a deficient organ*". This last definition integrates the biological environment – living system, biological fluids, proteins, cells and tissues) as well as the biological constraints and functions. Therefore biomaterials are different from other materials in the sense that they must have the capacity to exist in contact with tissues from the living system without causing an unacceptable degree of defence or a hostile host response.

2.2.4. *Extensions of the biomaterials field*

Then, despite the fact that the concept of biomaterials may appear as being quite complex with various co-existing definitions or meanings, the aim or the purpose is unique and well defined: *a biomaterial it is a material used and adapted for biomedical applications*.

During the last 40–50 last years, thousands of researchers, academics, surgeons of all specialties, regulatory agencies and companies have been involved and have endeavored in this interdisciplinary and transdisciplinary domain to make "biomaterials" a scientific discipline.

As seen in Chapter 1, biomaterials were for a very long time mainly devoted to quite simple applications which were more cosmetic than functional. Then the topic became more sophisticated when surgeons together with scientists took an interest in the field and proposed functional materials usable in dental, ophthalmologic, cardiovascular and orthopaedic surgeries. The first materials were very simple metals and/or alloys chosen for their mechanical properties and resistance to corrosion but during the first part of the 20th Century, the amazing development of the plastics industry as well the increased interest in the ceramic domain opened the way for new materials, endowed with physicochemical and mechanical properties as diverse as they were interesting. "Polymers" or plastic materials, ceramics and metals will be detailed further in the following chapters. Biomaterials used diversified and the science of biomaterials was born. Biomaterials were first "classified" as a category of materials, needing specific properties for one biomedical application. Today the applications are multiple: implants and prostheses for all surgeries, biodegradable scaffolds for tissue engineering, cell therapy and regenerative medicine, nanotechnologies, drug delivery mechanisms. Therefore biomaterials used for medical applications, engendered increasing derivatives; they are also used to grow cells, to control protein adsorption (purification, separation, competition…), to deliver therapeutic agents, as biosensors etc.

The evolution of biomaterials science tended to propose a "third generation" of biomaterial issued from new research areas with the aim of proposing better solutions to replace deficient organs and prevent the hostile host responses encountered when materials are implanted for the long term. Among these solutions, we find:

– the synthesis or the elaboration of new polymer, metallic or ceramic materials arising from the most recent research in macromolecular chemistry, metallurgy or inorganic chemistry. They represent a new generation of materials capable of offering properties which are very specific and/or even not yet known;

– the surface modification of materials still used as biomaterial implants or prostheses in order to improve their biological properties and induced host response – adsorption of proteins and physiologic fluids, cells response, surrounding tissues integration. There are several techniques to modify surfaces: coating, deep-coating, physisorption, chemical grafting. The purpose is to bind a molecule of biological interest to the surface of an implant made of a polymeric, metallic or ceramic material to induce a better hydrophobic/hydrophilic balance (make hydrophilic a hydrophobic surface), specificity (lead a perfectly controlled and precise biologic response) or bioactivity (modulate the host response);

– the synthesis and/or elaboration of "biodegradable" polymeric or ceramic structures. These three dimensional and porous structures also called "scaffolds" are intended to serve as a temporary support for cell attachment, growth and differentiation with the aim of creating a new functional tissue to replace one defect of a soft or hard tissue (bone, skin, ligament, etc.). The new tissue has to be able to restore the initial function of the replaced tissue and the scaffold is intended to be progressively degraded. This is tissue engineering, it is a part of biomaterials but in the peculiarity to necessarily involve materials and cells;

– last but not the least, the nanomaterials are the most recent materials used in the biomedical field. They belong to the biomaterials field which includes all the objects useful for medical application from the nanometric to the mesometric scale: they may be intended to be used in order to coat implant or prosthesis surfaces but besides this classic biomaterial application, nanomaterials are also used as "theranostic tools" for *in vitro* and *in vivo* diagnostics and imaging as well as in therapy through different new and effective drug delivery systems.

Because all the biomaterials are intended to be placed in contact with biological species for short or long periods, then have to respond to defined properties which can be summarized by the word "biocompatibility" which will be developed in Chapter 4.

2.3. Biomedical device

2.3.1. *Introduction*

What differentiates a biomaterial from a medical device? In medical applications, biomaterials are rarely used simply as a material and are more commonly integrated into more or less complex devices considered as medical devices. Similar to biomaterials, the identification of this sector is quite complicated because medical devices make up a vast and heterogeneous domain and this is one of the reasons why the industry of medical devices is often wrongly considered as being part of the medical industry domain. As we are going to see, the role of medical devices is closely linked to that of biomaterials. Indeed, and despite this misunderstanding of their importance, medical devices have well-identified purposes:

– the diagnosis, the prevention, the control, the treatment of a disease;

– the diagnosis, the control, the treatment, the compensation of a wound or a handicap;

– the study, the replacement or the modification of the anatomy or of the physiological process.

2.3.2. *Definition of a medical device*

Medical devices are products of health, including software, claiming a use in medical purposes and from which the mode of the deliberate main action is not obtained by pharmacological or immunological means nor by metabolism, otherwise the product qualifies as medicine (Article R5211-2 of the French Public Health Code).

Within this definition it is easy to understand how and why medical devices cover a very large domain of health engineering with applications as different as diagnostics, therapeutics, disease follow ups or handicap compensation. Medical devices are characterized by: (1) a very large diversity of products going from the simple patch to the most sophisticated cardiac valve, from wheelchairs to hip prostheses, from syringe or catheter to a vascular prosthesis, all the materials of diagnosis to heavy equipment of medical imaging; (2) a high number of referenced products (~1 to 2 million in France, IGAS report 2010) have been developed in response to the targeted needs of many small groups of patients (e.g. aortic endo-prostheses only a few thousand patients in France).

The market of medical devices holds an increasing place both at the world and European levels and represents a market of several hundred billion euros. Within this market, France takes fourth or fifth place with 10% of the world market. It is characterized by its industrial network made of 94% of small and medium sized companies (less than 250 people) among which 45% of very small sized companies (less than 20 people) and by numerous markets "niche" (report PIPAME 2011, France)

2.3.3. *Classes of medical devices*

If we come back to the above definition of biomaterials given by the experts as being "any material, natural or not, including all or part of a living structure or a biomedical device which executes or replaces a bodily function", it is evidenced that biomaterials dedicated to the implantation such as implants, prostheses, scaffolds of tissue engineering are implantable medical devices and then, belonging to this family of device, they are also distributed in four classes which are differentiated by the risk incurred by the patients : class I, class II a, class II b et class III (Article R5211-7 of the French Public Health Code), plus the class of active medical devices.

The classification and rules for medical devices are based on the degree of risk encountered by the patient. The classification takes several criteria into account such as the invasiveness of the device, the duration of its contact with patients, the part of the body which can be affected by its use. In Europe, all the details of the classification can be found in the Directive 93/42/EEC whereas in the United States the criteria are defined by the FDA etc. Whatever the country, the higher the class, the higher the potential risk for the patient and then the higher the number of tests to be passed before approval.

– The class I/which corresponds to low degree of risk – this class contains non-invasive devices or short-term use invasive devices in contact with skin (wheelchairs, support stockings, bandages used as mechanical barriers or to compress or absorb exudates, reusable surgical instruments, scalpels, etc.);

– The class IIa/medium degree of risk – this class contains invasive devices of short-term use (contact lenses, echographs, cutaneous staples, dental crowns, devices of long-term cell or tissue preservation, blood bags, etc.);

– The class II b/high degree of risk – contains invasive devices of long-term use (surgical implants, haemodialysis equipment, infusion pumps, etc.);

– The class III/highest degree of risk – contains invasive devices of short- or long-term used in contact with heart or central circulatory (CCS)/nervous system (NCS) (stent, hip prosthesis), or devices having a biological effect (administrator of medicine) or undergoing chemical changes in body (biodegradable device). Class III devices require premarket approval and preclinical and clinical assessments.

In addition to this classification, eighteen rules are edited in order to help manufacturers determine the degree of risk provide by their medical devices. In Europe, the safety of medical devices is assured by the "EC labeling" of the European Community under the control of National Agencies.

2.4. Other definitions: implant, prosthesis, organ, graft, etc.

Numerous terms are used when we speak about biomaterials. Having defined above in an extensive way biomaterials and medical devices, here is some essential terms frequently used in the biomaterials domain having subtle differences. The aim is to provide terminology usable without confusion. This is necessary because (1) human health and environmental sustainability are more and more interdependent, (2) research, applications, norms, and regulations are still developed independently in each sector, and (3) non-specialists like journalists, politicians, and partners of complementary disciplines are more and more implicated and need a common language.

Implant: An implant is a medical device (apparatus, delivery device, prosthesis, organ etc.) made of one or several biomaterials, which is introduced into the human body in the long term to replace an organ or to supply a function or to treat a disease. It can be implanted inside the body or under the skin (subcutaneous implant) or under an epithelial surface. The implant can be temporary (contraceptive implant) or permanent (dental implant, intra ocular lens).

Prosthesis: a device implanted in the body to supply a missing organ (limb, organ, tissue) or to restore a deficient function.

Artificial organ: is a device made of one or several biomaterials, which replaces a part, or all the functions of an organ.

Organ: part of the body of a human being performing particular functions.

Hybrid artificial organ: an artificial organ made of one or several biomaterials combined with living cells. The aim is to implant a new "functional organ", therefore the living cells have to be able to differentiate to form a new "functional tissue".

Bioprosthesis: an implantable prosthesis which is totally or substantially constituted of a "treated" non-living tissue, obtained from a donor.

Graft: set of living cells, living tissue or living organ surgically inserted into a body to replace a damaged part or a defect of an organ (transferred from a donor site to a receiving site). This is an autograft if the donor and recipient is the same individual; this is an allograft when the donor and recipient belong to the same species but are genetically distinct; this is a xenograft when the donor and recipient are of different species.

2.5. Tissue engineering, regenerative medicine, nanomedicine

The fourth generation of biomaterials is constituted of materials involved in the recent results from advances in biomaterials field: tissue engineering, nanomedicine and regenerative medicine.

2.5.1. *Tissue engineering*

Tissue engineering began more than twenty years ago but is still evolving. It is a big and major domain of biomaterials science. The goal of tissue engineering is not so different to that of biomaterials which is "to supply or to replace all or a part of a deficient organ". In both cases a material will be used but the approaches are different: (1) in one case (biomaterial) the deficient organ is replaced by a non-degradable material which is implanted for the long term and will stay until it will not work or break or cause damage, (2) in the other case (tissue engineering) the deficient organ or part of an organ is replaced by a three-component structure or "active construct" made of a biodegradable scaffold + living cells + bioactive factors the aim of which being the regeneration of a new functional tissue or organ at the same time as the scaffold is degraded. The major benefit of the tissue engineering approach is that the introduced implant material sensed as a foreign body is (bio)eliminated and replaced by a natural tissue. The principle is quite simple and very attractive, the results are not so easy to get and success is not guaranteed and depends on different parameters as:

– the nature, architecture and mechanical properties of the scaffold;

– the size of the defect;

– the nature of the tissue to be replaced.

Therefore, advances depend on the optimization of critical constituents: the scaffold, the participating cells and the regulation of bioactive factors. The goal here is not to describe and develop tissue engineering, is it just to present some basic principles. Many books, journals and conventions are entirely devoted to tissue engineering research and are good references for people who want to investigate tissue engineering research further.

Tissue engineering is based on principles and methods of engineering and life science; then, molecular biologists, biochemists, bioengineers, engineers, physical therapists and clinicians have to work together for solving these extremely difficult problems: the structure–function relationships of normal tissues have to be maintained with the "new tissue".

The scaffold: The ideal biomaterial to be used as a scaffold has to allow the adhesion, growth, differentiation and colonization of cells to get a new tissue which has the same functionality as the replaced one. It is generally a porous biomaterial. The porosity (size, distribution, interconnections diameters) plays a major role in the cell growth and rehabilitation: the greater the vascularization of the targeted organ the higher the developed porous volume and specific area. Many other parameters such as the physicochemical and mechanical properties of the material are very important: rugosity, surface chemistry, surface energy, class of material (polymer or ceramic), degradation and degradation products etc.

The choice of the cells and of the biological factors as growth factors is as important as the choice of the scaffold. Cells must have the capacity to keep and maintain its phenotype under varied circumstances. Cells may be autologous or not, differentiated or progenitors.

Two ways exist: first way consists *in vitro* of generating a functional tissue before its implantation, the second way is based on the *in vivo* implantation of an immature graft previously *in vitro* grown which will go on its maturation *in situ*.

2.5.2. *Regenerative medicine*

Regenerative medicine is quite a new discipline closley linked to tissue engineering and biology. It is considered as a branch of translational research in tissue engineering and molecular biology. The aim is common to that of tissue engineering with a process of replacing, engineering or regenerating human cells, tissues or organs to restore or establish normal function of damaged tissues or organs. The originality of this new field of research comes from the conditions of stimulation of cells and tissues using the body's own repair mechanisms to obtain new functional tissues or organs. Research in regenerative medicine is rapidly expanding whereas the regulation rules – uses and restrictions – are not yet finalized.

2.5.3. *Nanomedicine*

To introduce nanomedicine it is necessary to introduce nanotechnologies. This allows understanding the benefits of the new technologies.

In a strict sense, nanotechnologies may involve the research and development at the “atomic, molecular or macromolecular scale” carried out to create or elaborate structures, devices and systems of which the size is between 1 to 100 nanometers (nm). In reality, nanotechnologies also concern elements of micrometer size but onto which the miniaturization confers new properties.

Nanomedicine: nanotechnologies allow proposing new therapeutic approaches in medicine and new mechanisms of activity of therapeutic agents and medicines. The medical domains concerned by nanotechnologies are various: chemotherapy, radiotherapy, surgery, medical imaging (sensors, RMI). The applications to medicine are innovative in terms of prevention, of diagnosis and of treatment.

Prevention: nanoparticles are today principally used as filters in dermal cosmetics or as surface coating to prevent infection.

Diagnosis *in vitro* and *in vivo*: bioassays, nanoarrays, biosensors, imaging, contrast agents, improve the sensitivity, the reliability and the rapidity of detection and of monitoring as well as the miniaturization of devices.

Treatment: nanomaterials improve the targeting of therapeutic agents, allow a precise therapeutic follow-up and a more precise drug or gene delivery, surface modification, and allow decreasing the doses therapeutic agents.

The benefits are numerous: diagnosis is more precise and less expensive, the detection threshold is improved and pathologies are detected earlier, imaging is more successful, targeting is more efficient and specific, surgery is less invasive, secondary effects are lowered. This explains why the medical specialities, which already use, or will use in future, nanotechnologies are mainly, the oncologists, the specialists of metabolic diseases, of infectious diseases, of cardiology and haematology, neurodegenerative disease.

This new domain is expanding and several generations of nanomaterials are in development and co-existing: the miniaturization of existing devices and the molecular manufacturing to construct nanometric objects. Nanomedicine is used as numerous as various nano-objects such as colloids, aerosols, coating, ceramics, transistors, nanoparticles, liposomes, quantum dots and delivery systems.

2.6. Bibliography

[ARC 14] ARCOS D., BOCCACCINI A.R., BOHNER M., *et al.*, "The relevance of biomaterials to the prevention and treatment of osteoporosis", *Acta Biomaterialia*, vol. 10, no. 5, pp. 1793–805, May 2014.

[BAQ 99] BAQUEY C., "Biomaterials science: prospects for the new millenium?", *Journal of Materials Science: Materials in Medicine*, vol. 10, no. 12, pp. 695–696, December 1999.

[BAQ 05] BAQUEY C., "Tissue engineering: a new space of scientific expression and technological development for health", *Pathologie Biologie (Paris)*, vol. 53, no. 10, pp. 569–570, December 2005.

[BRY 12] BRYERS J.D., GIACHELLI C.M., RATNER B.D., "Engineering biomaterials to integrate and heal: the biocompatibility paradigm shifts", *Biotechnology and Bioengineering*, vol. 109, no. 8, pp. 1898–1911, August 2012.

[BRY 14] BRYSKIN A.V., BROWN A.C., BAKSH M.M., *et al.*, "Learning from nature – Novel synthetic biology approaches for biomaterial design", *Acta Biomaterialia*, vol. 10, pp. 1761–1769, 2014.

[LEN 12] LENDLEIN A., PIERCE B.F., AMBROSIO L., *et al.*, "Advanced functional polymers for medicine: multifunctional biomaterials", *Acta Biomaterialia*, vol. 8, no. 12, p. 4199, December, 2012.

[MIG 09] MIGONNEY V., HELARY G., "Bioactive polymer coating to improve bone repair", in PLANELL J.A., BEST S., LACROIX D. (eds.), *Bone Repair Biomaterials*, Woodhead Publishing, pp. 309–323, 2009.

[PAR 05] PARIENTE J.L., BORDENAVE L., CONORT P., "Biomatériaux et génie tissulaire en urologie – Partie A: aspects généraux – Chapitre III: Biomatériaux, Biomatériels, Biocompatibilité", *Progrès en Urologie*, vol. 15, no. 5, pp. 887–890, 2005.

[PLA 11] PFAFFENROTH C, WINKEL A., DEMPWOLF W., *et al.*, "Self-assembled antimicrobial and biocompatible copolymer films on titanium", *Macromolecular Bioscience*, vol. 11, no. 11, pp. 1515–25, November 10, 2011.

[RAT 87] RATNER B.D., JOHNSTON A.B., LENK T.J., "Biomaterial surfaces", *J Biomed Mater Res.*, vol. 21 (A1 Suppl.), pp. 59–89, April 1987.

[RAT 04] RATNER B., HOFFMAN A., SCHOEN F., *et al.*, *Biomaterials Science: An Introduction to Materials in Medicine*, 2nd ed., Elsevier Academic Press, 2004.

[REI 11] REICHERT W.M., RATNER B.D., ANDERSON J., *et al.*, "2010 Panel on the biomaterials grand challenges", *Journal of Biomedical Materials Research Part A.*, vol. 96, no. 2, pp. 275–287, February 2011.

[RIC 12] RICHARDS R.G., MORIARTY T.F., MICLAU T., *et al.*, "Advances in biomaterials and surface technologies", *Journal of Orthopaedic Trauma*, vol. 26, no. 12, pp. 703–707, December 2012.

[SER 14] SERRA T., ORTIZ-HERNANDEZ M., ENGEL E., *et al.*, "Relevance of PEG in PLA-based blends for tissue engineering 3D-printed scaffolds", *Materials Science and Engineering C: Materials for Biological Applications*, vol. 38, pp. 55–62, May 1, 2014.

[TUN 14] TUNG J.C., PAIGE S.L., RATNER B.D., *et al.*, "Engineered biomaterials control differentiation and proliferation of human-embryonic-stem-cell-derived cardiomyocytes via timed notch activation", *Stem Cell Reports*, vol. 2, no. 3, pp. 271–281, 2014 February 27.

[VER 12] VERT M., DOI Y., HELLWICH K.H., *et al.*, "Terminology for biorelated polymers and applications (IUPAC recommendations 2012)", *Pure and Applied Chemistry*, vol. 84, no. 2, pp. 377–410, 2012.

[WIL 99] WILLIAMS D., *The Williams Dictionary of Biomaterials*, Liverpool University Press, Liverpool, 1999.

[WIL 08] WILLIAMS D., "On the mechanism of biocompatibility", *Biomaterials*, vol. 29, pp. 2941–2953, 2008.

3

Materials Used in Biomaterial Applications

3.1. Introduction

Biomaterials are commonly used in various medical applications such as cardiovascular, gastrointestinal, urological, orthopedic, dental, plastic surgery, wound healing, tissue engineering and ophthalmology. Since the needs in biomaterials increase with the aging population and the desire to maintain health and well-being, the global biomaterial market is expected to reach $88.4 billion by 2017 with a compounded annual growth rate (CAGR) of 15% [MAR 13]. The major part of biomaterials and medical devices is composed of synthetic materials which are divided into three main classes: metals, bioceramics and polymers. The other classes are composites and biologically derived materials. Polymers represent more than half of the market, metals represent around a third, while ceramics only count for approximately 5% [SAE 99]. However, due to the development of biodegradable and biocompatible polymeric biomaterials, the biomaterial polymer market is expected to increase by 2017 with the highest CAGR of 22.1% [MAR 13]. It is clear that the economic and medical impacts of biomaterials highlight the importance of the research and development at the material scale.

Chapter written by Géraldine ROHMAN.

The design of implants and prostheses is a challenging process since the targeted materials must imitate the structure and properties of biological tissues and respond to specific requirements. These requirements are categorized into three classes [LON 98, NAG 12]:

– compatibility: the biomaterial must be non-toxic, non-allergenic, non-thrombogenic, non-antigenic and non-carcinogenic. The biomaterial can be inert or tissue activating, but it must not lead to local deleterious changes and cause little or no foreign-body reaction. Debris generation has to be minimized;

– mechanical and physical properties: the biomaterial should possess optimized properties depending on the application, such as elasticity, yield stress, ductility, toughness, ultimate strength, fatigue strength, hardness, wear resistance and porosity. Moreover, the biomaterial must resist to mechanical, biochemical and chemical constraints induced by the human body;

– manufacturing: the medical device should be relatively easy to fabricate with a high reproducibility and at reduced production costs. The fabrication process must keep the quality of the raw materials and give an excellent surface finishing or texture. The final material must be safely and efficiently sterilized.

Above all, the selection of the material itself is the main factor to form the medical device. All materials are constituted of atoms that are bonded together by interactions and all material properties may be attributed to the structural features on the atomic/molecular level [PAR 07]. In metals, metallic atoms are closely packed in a crystal structure and atoms are held together through a non-directional strong metallic bond which is produced by the electrostatic interactions between free electrons and metallic cations. This property provides the ability of metals to easily transmit electricity and undergo plastic deformation [DAV 03]. Ceramics are divided into two different groups based on the interactions between atoms. In ceramics like diamond and graphite, atoms are arranged in a crystal structure and share their valence electrons to form a covalent bond which is directional and strong. In ceramics containing metallic and non-metallic atoms like oxide, the solid structure is maintained by strong directional ionic bonds. This bond is produced by electron exchanges between atoms, and electrostatic interactions between metallic cations and non-metallic anions. Moreover, the structural atomic arrangements are limited due to the repulsive forces between like ions. The property of the covalent and ionic bonds provides ceramics with high hardness, high compressive strength and chemical inertness [TUR 09]. Polymers are macromolecules with a large number of repeat units constituted of covalently bonded atoms. If the repeat unit structure is regular enough, macromolecules may organize themselves into a crystal lattice leading to semi-crystalline polymers; otherwise, polymers are amorphous solids. The solid structural state of polymers is mainly due to weak secondary interactions between macromolecules, called Van der Waals forces. Hydrogen bonds may also arise if macromolecules contain hydrogen atoms covalently bonded to an electronegative atom such as oxygen, nitrogen and fluorine. The weakness of the secondary interactions provides polymers with mechanical properties and melting temperature inferior to metals and ceramics [DAV 03].

To help in selecting materials, mechanical property maps, known as Ashby maps, are used to compare a large amount of physical properties from various material groups as well as biological systems [ASH 89, MEY 08]. The map presenting Young's modulus as a function of strength of various synthetic biomaterials and biological tissues is given in Figure 3.1. It is possible to point out that there is a broad range in Young's moduli from 0.001 to almost 500 GPa and that the range of strength is also large from 0.3 to 5,000 MPa. Soft tissues, such as skin, cartilage, ligament, breast and vascular system, have strength and moduli below 100 MPa, and therefore it seems obvious to replace them with polymers that possess similar properties. In contrast, hard tissues such as bone have a slightly higher strength compared to soft tissues and possess a modulus higher by a factor of 10. As a consequence, metals and ceramics are used to replace bone [SIL 94]. As described earlier, other requirements are also taken into account to restrain the selection in each material class. A medical device can be made from one material type or combination of various components. For example, hip prosthesis is constituted of three different parts made from different materials (Figure 3.2.): the femoral stem and the outer acetabular cup are metallic; the inner acetabular cup may be in polymer or ceramic and contacts a metallic or ceramic femoral head. Indeed, each material is used for its own properties: metals are needed when the device endures high mechanical loads; bioinert ceramics or high-strength polymers possess a wear resistance that limits debris generation from pieces which undergo friction; bioactive ceramics are used in coating to form direct chemical bonds with bones and ensure the implant stability without the use of cements [PAT 12].

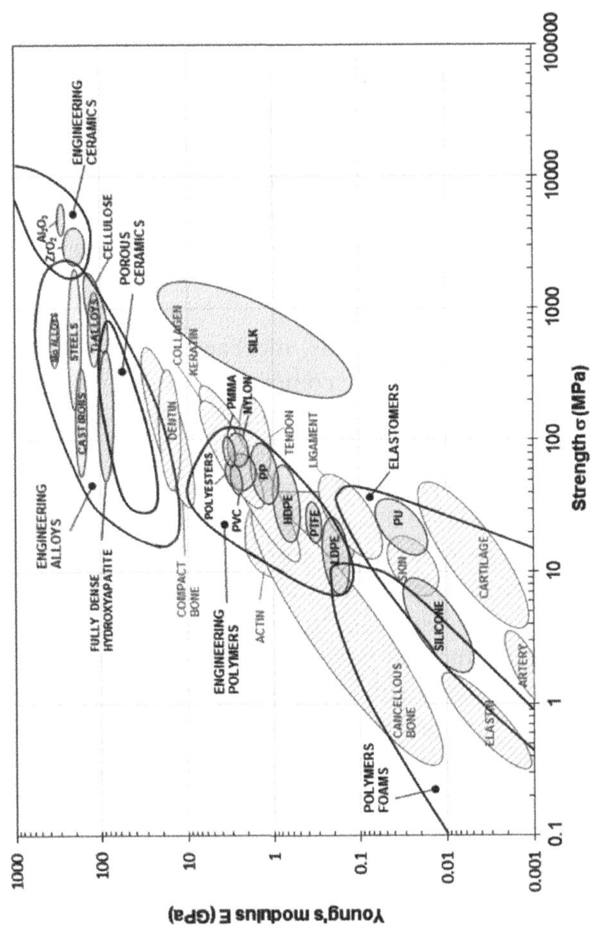

Figure 3.1. *Young's modulus versus strength materials selection chart (data from [ASH 89, KNO 11])*

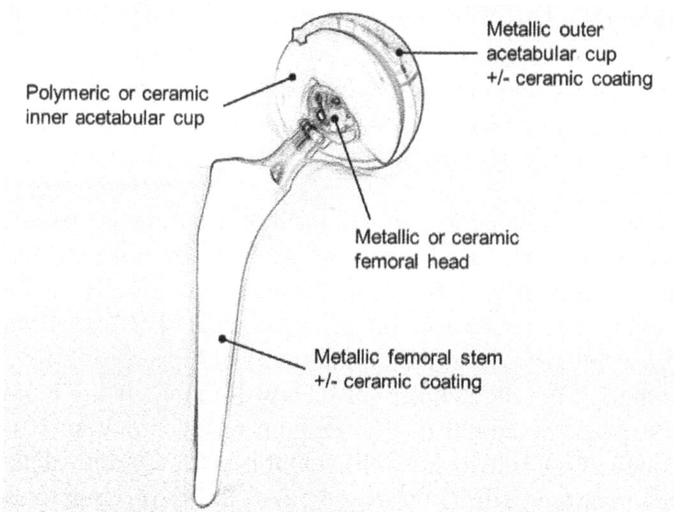

Figure 3.2. *Materials used to design a hip prosthesis*

In this chapter, each class of materials that are used as biomaterials is considered. Details on metals, ceramics and polymers applications as well as their mechanical properties will be given. Comments on the surface of each material type and their modifications will be made.

3.2. Metals and alloys

Metals have a high wear resistance and are strong and ductile, which make them appropriate for bearing large loads without leading to large deformations and permanent dimensional changes. They are mainly used to replace or fixe bone and dentin tissues. As a consequence, metals are used in a wide range of medical device applications such as fracture fixation (bone plates and screws), joint replacement, and dental appliances (braces, dental root implants). However, metals are dense, can be difficult to make and may corrode [PAR 07]. Furthermore, it is difficult to make a generic comparison between metals because of the relationship between mechanical design and selection materials. Indeed, their mechanical properties are mainly controlled by the material microstructure, that is to say the size of the grains: stronger materials are obtained with a fine-grained structure. Fatigue and yield strengths vary with the metal or alloy type and the processing, while Young's modulus is mostly set by the material type [CAR 04, DAV 03].

The choice of metallic atoms constituting alloys is based on the non-toxicity. In fact, metal ions may be released from the materials into the surrounding tissue and concentrate locally or diffuse systemically. The amount of the released ionic species and the type are of great importance to define the material biocompatibility. Apart from titanium (Ti), the majority of metals used in existing biomedical alloys, such as vanadium (V), aluminum (Al), chromium (Cr), iron (Fe), cobalt (Co) and nickel (Ni), exhibit some degree of negative biological impact [BIE 12]. Nevertheless, the most commonly used metals are titanium-based alloys, stainless steels and cobalt–chromium alloys. Other metals and alloys include commercially pure titanium (Ti-Cp), nitinol (shape-memory alloys (SMAs) based on nickel and titanium) and tantalum. Their main characteristics are given in Table 3.1.

3.2.1. *Titanium and titanium-based alloys*

Pure titanium crystallizes in hexagonal close-packed structure (α-phase) at temperatures below 883°C and in body-centered cubic structure (β-phase) at higher temperature. Addition of other elements leads to alloys with different structures: for example, Al, O, N, and C are α-stabilizers, while Fe, Mo, V, Co and Mn are β-stabilizers. Titanium and its alloys possess a stable passive oxide layer (mainly TiO_2) at their surface. They have a very good biocompatibility, corrosion and fatigue resistance and a relatively low modulus in comparison with stainless steel and Co-Cr-Mo alloys. However, due to the high processing cost, titanium is expensive. Moreover, Ti and its alloys have very poor wear resistance, which makes them unsuitable for load-bearing articulating surfaces [PIL 09].

Material	Atomic Composition	Density ($g.cm^{-3}$)	Tensile Strength (MPa)	Elastic Modulus (GPa)	Elongation (%)	Applications
Ti-Cp	Ti balancing and max. 0.5% Fe, 0.40% O, 0.1% C, 0.05% N, 0.015% H	4.5	240–550	100	15–25	pacemaker cases, dental implants, screws and staples for spinal surgery
Ti-6Al-4V	Ti balancing, 5.5-6.5% Al, 3.5-4.5% V, 0.25% Fe, and max. 0.13% O, 0.08% C, 0.05% N	4.4	860–985	110	10–15	Joint components, screws
Nitinol	Ti balancing, 55-56% Ni, and max. 0.76% Cr, 0.66% Fe, 0.64 Co%, 0.64% Mn	6.5	690–1380	≥48	13–40	Stents, medical staples, shape memory plates, guidewires
Co-Cr-Mo Alloys	Co balancing, 27-30% Cr, 5-7% Mo, 1% Ni, 0.75% Fe, 0.35% C and max. 1% Mn, 1% Si, 0.3% Al, 0.25% N	8.3	655–1277	230	10–30	Fracture plates, heart valves, joint components, screws, dentistry castings
Stainless Steel 316L	Fe balancing, 17-19% Cr, 13-15% Ni, 2.25-3% Mo, and max. 2% Mn, 0.75% Si, 0.5% Cu, 0.1% N,	7.9	490–1350	200	5–40	Fracture plates, pins, sutures, wires, total joint prosthesis (in UK)
Tantalum	Ta balancing and max. 0.1% Nb, 0.05% W, 0.02% Mo, 0.015% O, 0.01% Ni, C, Fe, Ti, N	16.6	285–650	190	5–30	Suture wires, clips for the ligation of vessels, staples, pliable sheets and plates

Table 3.1. *Main properties and applications of various metallic biomaterials (data from [DAV 03, MOR 04, ONG 14])*

Ti-Cp is available in four grades, depending on the content of oxygen. Grade I with the lowest content (0.18% max.) has the lowest yield strength and the highest ductility. In contrast, grade IV with the highest content (0.40% max.) has the highest strength and the lowest ductility (Table 3.1) [PIL 09]. Ti-Cp is widely used for dental implants because it promotes osseointegration. Indeed, it seems that the hydroxylated surface of Ti-Cp reacts with bone mineral phase constituents.

Ti6Al4V is the main Ti alloy used in load-bearing medical applications. The main alloying elements are aluminum (5.5–6.5%) and vanadium (3.5–4.5%), and thus it is an α + β Ti alloy. These alloys are used to obtain biomaterials with higher strengths than Ti-Cp (Table 3.1.) but with the same corrosion resistance and osseointegration properties. The fatigue strength of these alloys depends strongly on the size and distribution of the α + β phase regions. Due to its excellent corrosion resistance, Ti6Al4V is also used as porous-coated or surface-textured orthopedic implants [PIL 09]. Since vanadium is considered carcinogenic and highly cytotoxic [BIE 12], other Ti alloys, such as Ti6Al17Nb and Ti5Al2.5Fe, are considered as recent alternatives. They have similar properties and applications as Ti6Al4V (Table 3.2). Finally, besides a high fatigue strength, near-β and β Ti alloys are developed to obtain materials with lower Young's modulus, allowing us to avoid stress shielding and promote new bone formation.

Alloy	Elastic modulus (GPa)	Tensile strength (MPa)	Elongation (%)
Ti6Al17Nb	114	869–1,086	7–16
Ti5Al2.5Fe	107–112	860	8
Ti15Mo5Zr3Al	80	882–1,177	15–20
Ti12Mo5Zr5Sn	<75	1,010	17.8
Ti30Nb	<75	700	20
Ti30Ta	<75	740	28

Table 3.2. *Mechanical properties of vanadium-free α + β Ti alloys, β and near-β Ti alloys (data from [BRE 98, MAR 05])*

3.2.2. *Stainless steel*

Stainless steel has high strength, chemical inertness and resistance to corrosion due to the presence of chromium (Cr) that forms a thin oxide film which resists oxidation (Cr_2O_3) (a minimum of 12% of Cr is necessary to prevent corrosion). In medical applications, typical medical grade of stainless steel is the austenitic 316L, which is predominantly iron alloyed with major amounts of chromium (17–19%) and nickel (13–15%) (Table 3.1). The "L" in 316L stands for low carbon content (0.03%). This low amount of carbon reduces carbide precipitation at grain boundaries ($Cr_{23}C_6$), thus the Cr depletion from adjacent zones, and therefore minimizes *in vivo* corrosion. Indeed, zones with reduction of Cr content are more susceptible to intergranular corrosion because of a less stable passive oxide film [PIL 09]. Since Cr tends to stabilize the weaker body-centered cubic ferritic phase, the presence of high nickel amounts in 316L counteracts this tendency and stabilizes the face-centered cubic austenitic phase of steel [DES 08]. The cost, the availability in various stock forms and the ease to process 316L make it easy to design a wide range of final implant shapes with a wide range of mechanical properties. However, stainless steels corrode under highly stressed and oxygen-depleted environments. Combined with their high modulus, they are mainly used for temporary devices [DES 08].

The other main disadvantage of 316L is the presence of nickel. Indeed, Ni has generally a poor biocompatibility: susceptibility to corrosion via biological fluids, high relative cytotoxicity, hemolytic behavior in particulate form, genotoxicity, carcinogenicity and potential mutagenicity [BIE 12, YAN 10]. As a consequence, nickel-free austenitic stainless steels are considered. BioDur®[1] 108 is an alloy with high nitrogen content which confers higher tensile and fatigue strength than 316L. Its resistance to corrosion is also superior. Its properties tend to be the following: density 7.7–8.03 g.cm^{-3}, elastic modulus 200 GPa, ultimate tensile strength 931–2, 206 MPa and 3–49% elongation. BioDur® 108 is used in bone plates, spinal fixation, screws, and hip and knee components [CAR 05].

3.2.3. *Cobalt-based alloys*

Cobalt-based alloys possess good wear resistance and fatigue strength. Thus, they are used as prosthesis stems, load-bearing components in joint replacement devices and dentistry castings. There are mainly two types of Co-based implants: Co-Cr-Mo casted alloys, and Co-Ni-Cr-Mo and Co-Cr-W-Ni alloys wrought by hot forging. Castable alloys are principally used in dentistry and in making joint components, while the applications of wrought alloys are the stems of

1 BioDur® 108 composition: Fe balancing, 21–24% Mn, 19–23% Cr, 0.5–1.5% Mo, min. 0.9% N and max. 0.75% Si, 0.25% Cu, 0.10% Ni, 0.08% C, 0.03% P, 0.01% S [CAR 05].

hip or knee prosthesis. The main disadvantages of Co-based alloys are high moduli and density, which are almost twice those of Ti-alloys, and this increase in modulus may have an effect on the load transfer to bones in orthopedic devices [PAR 07].

The most widely used Co-based alloy in medical applications is the Co-Cr-Mo alloy known as Vitallium (Table 3.1). The good corrosion resistance of this alloy is attributed to the high chromium content (27–30%) which produces a Cr_2O_3 passive film at the implant surface. The structure of the alloy is a metastable face-centered cubic austenite (Co-rich matrix) with interdendritic regions rich in Cr, Mo and C that form carbides (primarily $Cr_{23}C_6$). For high-C alloy (0.35%), the carbide regions are hardened during functional loading, and therefore the cored structure is responsible for the material's high wear resistance. Indeed, Co-Cr-Mo alloys are the most resistant metallic biomaterials [PIL 09]. However, the carbide zones are sites for crack initiation and may define propagation pathways along the grain boundaries. In addition, Co-based alloys are difficult to machine due to a low ductility resulting from the extensive carbide networks. As a consequence, they are generally casted leading to coarse grain size which gives weaker materials in comparison with wrought alloys [DAV 03]. To reduce the grain size and obtain a higher strength, molybdenum is added (5–7%). Moreover, to design the casted alloy, a ceramic mold is used and the mold temperature impacts the grain size (from hundreds of microns to the millimeter): high temperature leads to larger grains and therefore inferior mechanical properties. Nevertheless, high-temperature process gives large and far apart carbide precipitates resulting in a less brittle material [PAR 07]. Finally, defects may arise from the casting due to ceramic inclusions in the metal. These kinds of inclusions would be a site for crack initiation and contribute to fatigue fracture of the device [BRU 04].

To avoid any problems that might be induced by casting processes, Co-Ni-Cr-Mo and Co-Cr-W-Ni alloys[2] were developed with almost

2 Co-Ni-Cr-Mo composition: Co balancing, 33–37% Ni, 19–21% Cr, 9–10.5% Mo, and max. 1% Ti, 1% Fe, 0.15% Si, 0.15% Mn, 0.025% C, 0.015% P, 0.01% S. Co-Cr-W-Ni composition: Co balancing, 19–21% Cr, 14–16% W, 9–11% Ni, 1–2% Mn, 0.05–0.15% C and max. 3% Fe, 0.4% Si, 0.04% P, 0.03% S [BRU 04].

the same moduli as Co-Cr-Mo alloys (210–230 GPa) [BRU 04]. To enhance fabricability, Cr content is reduced while Ni and W are added. To produce wrought Co-based alloys, the content of carbon is reduced, leading to a decrease in carbide precipitation and therefore a lower strength. However, in Co-Ni-Cr-Mo alloys, the high content of nickel (33–37%) stabilizes the hexagonal close-packed phase (hcp) and hcp bands emerge in the face-centered cubic grains, which results in a strengthened structure. Moreover, to maintain a good corrosion resistance and counteract the decrease in Cr content, the content of Mo is enhanced. This also strengthens the material because of the precipitation of Co_3Mo within the hcp phase. Co-Ni-Cr-Mo alloys have eventually a superior fatigue (500–793 MPa) and ultimate tensile strength (1,206–1,795 MPa) than the Co-Cr-Mo alloys, and they are the strongest alloys available for medical applications. They are particularly suitable for long service life device such as the stems of hip implants [BRU 04]. Co-Cr-W-Ni alloys are not as corrosion resistant as alloys containing Mo. They are mainly used for fracture fixation implants [PIL 09]. Due to the cost of tungsten and cobalt, Co-Cr-W-Ni alloys are significantly more expensive than stainless steel, which limits their usage [ONG 14]. Finally, some concerns arise with these alloys due to the presence of poor biocompatible Ni. However, the rate of nickel ion release was found to be the same for Co-Ni-Cr-Mo and 316L stainless steel despite a higher Ni content in the Co-based alloy [PAR 07].

3.2.4. *Shape–memory alloys*

SMAs are materials that respond to stress or heating by undergoing transition in their metallic crystal structure. Indeed, the martensitic phase is a low-temperature stable phase which is easy to deform, while the austenitic phase is stable at high temperature and is a rigid body-centered cubic arrangement. The properties of SMA are determined by the atomic composition and the processing methods. One-way SMAs are materials that undergo deformations at low temperature (in the martensitic phase) and return to their original shape when heated to their austenitic phase. Two-way SMAs are more complex since they are materials recovering a specific shape when heated and finally returning to an alternate shape when cooled to the martensitic phase. Another unique property of SMA is the superelasticity or pseudo-elasticity. In this case, the material is deformed at a constant temperature in the austenitic phase leading to a transformation into a martensitic phase. As soon as the load is removed, the martensite reverses to austenite bringing the material back to its original shape [BAR 00, DES 08]. In a superelastic material, 8% strain may be recoverable [ONG 14].

Nitinol, a Ni–Ti alloy (55–56% Ni, 44–45% Ti – Table 3.1), is a pseudo-elastic alloy used in medical applications since it exhibits the shape–memory effect near room temperature. Nitinol is used in orthodontic dental wires, medical staples and vascular stents [DUE 99, MOR 04]. Nitinol exhibits a ductility comparable to most ordinary alloys, which allows manufacturing in various forms, and a good corrosion resistance due to the TiO_2 passive surface layer like in other titanium implants. The alloy has a relatively low elastic modulus and is more resilient than stainless steel or Co-Cr-based alloys [PAR 07]. Moreover, despite high Ni content, nitinol has a good biocompatibility. As a matter of fact, the corrosion susceptibility of nickel is mitigated by the TiO_2 protective layer which limits the release of nickel ions and therefore the risk of cytotoxicity [BIE 12].

3.2.5. *Tantalum*

Tantalum (Ta) is one of the most biocompatible metals because of its low cytotoxicity and excellent corrosion resistance due to the stable Ta_2O_5 oxide formed on the material surface [BIE 12]. Ta can be used as an alloying element in Ti-alloys as well as in its commercially pure form (Table 3.1). Ta is a refractory metal with a high melting temperature (3017°C). Its structure is a body-centered cubic α phase which makes Ta very hard, and very ductile. However, due to its mechanical properties and high density, Ta is restricted to a few applications. It is successfully used in suture wires, clips and staples [PAR 07]. Porous Ta has also been tested as bone graft substitutes and cementless components in hip and knee arthroplasty considering its resemblance to cancellous bone in terms of porosity (75–85%) and stiffness. Its low stiffness (elastic modulus of 2.5–3.9 GPa) and high friction coefficient (from 0.82 to 1.75) reduce stress shielding and allow immediate weight-bearing. Ta also has an osseocompatibility similar to Ti-based devices. Moreover, when alkali-treated, a hydroxylated surface is created that may react, like Ti-based implants, with bone mineral phase constituents and allows the binding of the Ta-based implant to bone [LEV 06].

3.2.6. *Surface coating and finishing*

In biomedical applications, it is necessary to control the material surface since it plays a major role in the biological response. The surface disadvantages of metallic implants are the absence of function that favors implant bioactivity, and the difference in material composition between the surface and the bulk. Indeed, metals have high surface energy and organic materials with lower energy may adsorb easily. This leads to an oxidized and contaminated layer over the metallic surface which is not appropriate as is for biomedical applications. Many surface modifications have been developed to counteract these problems and provide specific surface properties [LIU 04]:

– mechanical finishing (polishing, grinding and blasting): cleaning and roughening to improve adhesion;

– chemical finishing (alkaline, acidic or hydrogen peroxide treatment, anodic oxidation): removing oxide scales and contamination, improving corrosion resistance, enhancing biocompatibility and bioactivity;

– physical finishing (physical vapor deposition, ion implantation and deposition): modifying surface composition, improving wear and corrosion resistance;

– chemical coating (modification and grafting through surface silanization, photochemistry, etc.): immobilizing bioactive polymers, peptides, proteins or growth factors, inducing specific cell and tissue response;

– physical coating (flame or plasma spray): improving wear resistance and corrosion, enhancing biological properties.

As already discussed, metals and alloys possess at their surfaces a stable oxide layer which protects them from corrosion. However, this thin layer is in the order of nanometers and may be destroyed by wear and fretting. If the oxide film remains unrepaired, corrosion arises and metallic ions release increases [HAN 99]. Thus, passivation is one of the most important surface modifications applied to metallic medical devices. Passivation enhances the protective oxide film by changing its composition, structure and thickness [BAL 08]. Coating of bioactive bioceramics, such as hydroxyapatite (HA), is a widely used surface modification to increase osseointegration of metallic implants. When the device is implanted in the body, ionic exchange appears between the ceramic surface and the physiological fluids, leading to the HA dissolution, precipitation of biological apatite and finally bone-binding [DEM 12]. Lately, surface modifications by immobilization of bioactive polymers or biomolecules have been developed to design biomimetic surfaces that allow guiding cell behavior [DUM 13]. Some examples of metallic surface modifications are given in Chapter 6.

3.3. Bioceramics

Ceramic materials are divided into two classes based on the nature of the atomic bindings: covalent and ionic bonds. Covalently bonded materials are carbonaceous structures and ionic-bonded materials are mainly oxide-based materials. Despite the existence of a wide range of ceramics, the choice of materials used in biomedical applications is reduced, and only a few were employed in human clinical applications. The main advantages of bioceramics are very good biocompatibility, good degradation resistance in corrosive environments, high compression strength and moduli, superior hardness and wear resistance in comparison with metals. These properties are the reason why bioceramics are used to repair or replace hard tissues such as bone or dentin [DAV 03]. However, ceramics have high melting temperatures, low heat conductivities, and they are difficult to shear plastically. As a consequence, they are difficult to fabricate and need specific process techniques, such as sintering which may lead to residual porosity [TUR 09]. Moreover, ceramics have values of tensile strength lower than the compressive ones, low impact resistance and, therefore, are brittle which limits their usage [PAR 07].

Depending on the nature of the tissue attachment mechanism, ceramics are divided into three subclasses: nearly bioinert ceramics, such as alumina, zirconia and pyrolytic carbon; bioactive ceramics, such as hydroxyapatite (HA), bioglasses or glass-ceramics; and bioresorbable ceramics, such as tricalcium phosphate (TCP). Their main characteristics and applications are given in Table 3.3.

Material	Chemical Composition	Density ($g.cm^{-3}$)	Compressive Strength (MPa)	Elastic Modulus (GPa)	Hardness (HV)	Applications
Alumina	Al_2O_3 (≥99.5%)	3.8-3.9	3000-5000	380-410	2000-3000	Heads and sockets of hip prostheses, femoral or tibial components of knee prostheses, dental implants, cochlear implants
Zirconia (Y-TZP)	ZrO_2 (≥93.2%) + 2-3% Y_2O_3	5.7-6.0	2000	195-210	1000-3000	Articulating ball in hip prostheses, load-bearing prostheses, dental implants
LTI Carbon		1.7-2.2	500-900	17-28	(236 DPH 500g)	Heart valves, hand joints, dental implants
Dense HA	$Ca_{10}(PO_4)_6(OH)_2$	3.2	100-900	70-120	500-800	Dental implants, alveolar ridge augmentations, maxillofacial reconstruction, coatings for chemical bonding
TCP	$Ca_3(PO_4)_2$	3.1	450-650	90-120		Temporary bone space fillers
45S5 Bioglass®	45% SiO_2, 24.5% CaO, 24.5% Na_2O, 6% P_2O_5	1.8-2.9	800-1200	40-140	458	maxillofacial prostheses, small bone replacement (middle ear surgery), coatings for chemical bonding
A-W Cerabone®	44.9% CaO, 34.2% SiO_2, 16.3% P_2O_5, 4.6% MgO, 0.5% CaF_2	3.1	1080	118	680	Bone defect fillings, vertebral prostheses, coatings for chemical bonding

Table 3.3. *Main properties and applications of various ceramic and glass biomaterials (data from [DAV 03, HAU 98, HEN 04, LI 98, THA 04])*

3.3.1. *Nearly bioinert oxide-based ceramics*

With nearly inert materials, the tissue is not chemically or biologically linked to the implant and the attachment occurs through morphological or biological fixations, that is to say through tissue ingrowth into surface irregularities and pores, respectively. For morphological fixations, movements at the interfacial zone may arise leading to biomaterial or tissue deterioration and fibrous capsule formation. The thickness of the fibrous tissue depends on material and the extent of implant motion. A thick fibrous capsule results in loosening of the medical device. In the case of biological fixations, the interfacial area increases, leading to the enhancement of the movement resistance. However, micromovements may damage tissue vascularization, leading to blood supply cutoff and tissue necrosis. Furthermore, the mechanical strength of the material decreases with the volume fraction of porosity [HEN 04].

Alumina and zirconia are the two main oxide-based nearly bioinert ceramics used as medical devices. Indeed, due to their structure, they are chemically stable and the release of substances to the surrounding tissue is very low. They have shown non-toxicity and a good biocompatibility. Moreover, alumina leads to a very thin capsule formation and prostheses may be fixed cementless when designed in a very tight fit [DAV 03].

Medical grade alumina-based devices are polycrystalline α-Al_2O_3 with fine grains (<4 μm) and a hexagonal close-packed structure. The material is chemically pure with very little amount of grain boundary impurities. Like with metallic biomaterials, attention has to be paid to the grain size since larger grains lead to a decrease in the material strength. The American Society for Testing and Materials (ASTM) standards for alumina-based implants require a flexural strength higher than 400 MPa and an elastic modulus of 380 GPa [PAR 07]. Due to its high purity and resistance to corrosion related to its phase stability, alumina has a long-term stability which contributes to its use as dental implants. Alumina possesses high mechanical strength and the highest hardness among oxide-based bioceramics (Table 3.3). Moreover, alumina-based devices have a high surface finishing with accurate dimensions, a low friction coefficient (0.044–0.115) and wear rate (1–5 mm^3/year *in vivo*) that make alumina a good candidate for joint replacement materials despite its brittleness. Nevertheless, its high elastic modulus may be responsible for bone atrophy and implant loosening in old patients. Finally, another limitation of alumina-based biomaterials is its high manufacturing cost [LI 98].

The main advantage of zirconia (ZrO_2) is its lower elastic modulus (Table 3.3). Moreover, zirconia can exist in different phases (monoclinic, tetragonal and cubic) which can be tailored to enhance material toughness and strength. Pure zirconia is monoclinic at room temperature, but the addition of other oxides, such as CaO, MgO and Y_2O_3, allows the generation of multiphasic materials. Furthermore, by stabilizing zirconia in cubic or tetragonal phases, the additives prevent from the material volume change and cracking that occurs during cooling down and phase transformation. Zirconia-based medical devices are divided into two categories: partially stabilized zirconia (PSZ) and tetragonal zirconia polycrystals (TZPs). PSZ is constituted of a cubic major phase and precipitates of monoclinic and tetragonal phases at grain boundaries or within the cubic matrix grains. Y-TZP materials contain 2–3% Y_2O_3 and are completely constituted of tetragonal grains in the order 0.4–0.8 µm. Both PSZ and Y-TZP have been used as medical implants, but Y-TPZ materials are the most selected into the market ball heads. Indeed, the flexural strength (950 MPa) and the fracture toughness (10.5 $MPa.m^{1/2}$) of Y-TPZ are almost the double of alumina-based ceramics, leading to a decrease in sensitiveness to stress concentration at contact points [PIC 99]. Moreover, in comparison to alumina, Y-TPZ shows lower friction coefficient (0.028–0.082), finer grain size and better-controlled microstructure without any residual porosity. As a consequence, Y-TPZs are good candidates to replace alumina in orthopedic applications or dental crowns. However, the main drawback of zirconia is that it may be weakened due to phase transformations that may arise under loading leading to surface degradation, or that occur through Zr-OH bonding when subjected to aqueous environment [PAR 07] Chapter 9.

3.3.2. *Carbon-based implants*

Carbon-based materials can be obtained in various allotropic forms; but only the low-temperature isotropic (LTI) and the ultralow-temperature isotropic (ULTI) pyrolytic carbons are widely used as biomaterials, mainly as surface coating onto articulating joint surfaces for ULTI carbon, and hand joints and heart valves for LTI carbon. Pyrolytic carbons are man-made pure elemental carbon materials obtained from the pyrolysis of hydrocarbon precursors. They are partially crystalline materials since they have a turbostratic structure which consists of graphene sheets held by Van der Waals forces and stacked in a disordered manner through random rotations or displacements of the layers relative to each other. Pyrolytic carbons exhibit small crystallites (2.5–4.0 nm for LTI and 0.8–1.5 nm for ULTI) randomly oriented conferring to the material its isotropic behavior [BOE 11].

High-purity pyrolytic carbons (Table 3.3) are bioinert materials that possess low modulus, quite high strength in comparison with glassy carbon and graphite (compressive strength = 172 and 138 MPa, respectively), high fatigue and wear resistance, good compatibility with blood and soft tissue. The mechanical properties are related to the density, that is to say to the material aggregate structure. High density LTI carbons are strong materials which can be designed as coating or as monoliths, while ULTI carbons can also be obtained with high density (1.5–2.2 $g.cm^{-3}$) and strength but only as a thin coating (0.1 to 1 μm). The combination of low modulus and high flexural strength (275–550 MPa for LTI and 345 to >690 MPa for ULTI) leads to large strain to failure (2% for LTI and >5% for ULTI). As a consequence, it is possible to coat flexible polymeric biomaterials since the coating does not fracture under flexion of the substrate [DAV 03]. Another unique property of pyrolytic carbons is their durability, and they do not fail in fatigue because of the absence of mobile defects in their crystalline structures. Furthermore, up to 20% of silicon can be added to LTI resulting in a structure composed of sub-micron β-SiC particles randomly dispersed in a matrix of roughly spherical micro-size subgrains of pyrolytic carbon. These silicon-alloyed LTI carbons are developed to improve stiffness, hardness, wear resistance and strength [RIT 96]. The main drawbacks of pyrolytic carbons are the lack of bone fixation leading to the loosening of joint prostheses, the apparition of cavitation and pitting due to blood flow and impact stress resulting from the implant failure, and despite their good hemocompatibility, the platelet adhesion on the implant surface which may induce thrombus formation [BOE 11].

3.3.3. *Bioactive ceramics*

Bioactive implants elicit a specific biological response at the material surface leading to the formation of a chemical bond between the material interface and the tissue. Some materials have demonstrated the ability to develop bioactive fixations to tissue with various mechanisms and rates of bonding, and different thicknesses and strengths of the interfacial zone. These materials include bioactive glasses and glass-ceramics, hydroxyapatite (HA) and some composites. They develop at their surface a layer of hydroxyl carbonate apatite which is chemically and structurally equivalent to the mineral phase of bone. Due to its equivalence, the layer can bond with collagen fibrils and thus to bone. The surface reaction is dependent on the material composition and small changes can totally suppress the bioactive property. In bioactive fixation, the adherent interface resists to high mechanical forces. Indeed, the adhesion strength of the interfacial zone is equivalent to or even greater than the cohesive strength of the only material or the only tissue. For example, when testing adhesion in a pull-off model, the failure of the device appears in the implant when using Bioglass® and in the bone when using glass-ceramic Cerabone®. In addition, the bioactive fixation is 15–40 times stronger than the fixation developed with a non-bioactive material (morphological fixation) like alumina where the failure of the device occurs in the interfacial zone [HEN 98].

Common inorganic glasses are developed from the SiO_2-CaO-Na_2O system. Bioactive glasses are single phase vitreous materials which are based on a stable vitreous silica-rich matrix (38–65% SiO_2) containing oxide in specific amounts, such as Na_2O (15–30%), CaO

(10–24%), P_2O_5 (0–8%), B_2O_3 (0–3%), Al_2O_3 (0–3%). The main advantage of CaO and P_2O_5 is obviously calcium and phosphorus, which are major constituents of the mineral phase of bone. CaO may be partially replaced by MgO or CaF_2, and Na_2O by K_2O, with little changes in bioactivity. Al_2O_3 and B_2O_3 may substitute SiO_2 to alter the glass production process or their surface dissolution rates, but they have to be added at low content to avoid the inhibition of the bioactivity [DEA 07]. Despite the development of numerous bioactive glasses, the original Bioglass® 45S5 has been found to have the best biological properties with a high bioactive index (12.5) (Table 3.3). The interface that forms between Bioglass® 45S5 and bone is thick (200 μm) but possesses a low shear strength. Moreover, the material has low mechanical strength and toughness, and is therefore used as coatings in particulate form or in low load-bearing applications, such as alveolar ridge maintenance or to replace the ossicular chain of the middle ear [HEN 98].

Bioactive glass-ceramics are polycrystalline materials obtained through appropriate thermal treatment of glass. They possess a fine and homogeneous grain size with few or no residual porosity, leading to improved mechanical properties (good mechanical strength and toughness). Bioactive glass-ceramics are based on the composition of Bioglass® with low amounts of alkali oxides and therefore have intermediate bioactivity index. One of the most bioactive glass-ceramics used in biomedical applications is Cerabone® A-W (Table 3.3). This material is constituted of two crystalline phases: a β-wollastonite ($CaO.SiO_2$) phase (34%) and a oxyfluorapatite ($Ca_{10}(PO_4)_6(O,F_2)$) phase (38%), plus a residual glass phase (28%). Both crystal phases are homogeneously distributed in the glass matrix with a grain size of 50–100 nm. The β-wollastonite phase acts as a reinforcing phase which prevents straight propagation of cracks. Cerabone® A-W has the highest mechanical strength among all bioactive glasses and glass-ceramics. Its fracture toughness (2.0 $MPa.m^{1/2}$) is nearly tripled in comparison with its parent glass (0.8 $MPa.m^{1/2}$) [WAN 04]. Its moderate bioactivity index (3.2) leads to a thin bonding interface (10–20 μm) which, however, possesses a high resistance to shear. Indeed, it is reported that a value of bioactivity index around 4 leads to an optimal interfacial bonding strength [HEN 04]. Cerabone® A-W flexural strength (220 MPa) is much higher than that of its parent glass (72 MPa), nearly the double of hydroxyapatite (115 MPa) and exceeds slightly that of human cortical bone (160 MPa) [DEA 07]. As a consequence, Cerabone® A-W is used in moderate load-bearing applications, such as iliac crest and vertebral prostheses, dental implants and maxillofacial reconstruction. The main drawbacks of glass-ceramic biomaterials are its brittleness (like other glasses and ceramics), the impossibility to substantially improve mechanical strength due to composition restriction as defined for a good biocompatibility [PAR 07].

Since bone and dentin tissues contain hydroxyapatite, synthetic polycrystalline hydroxyapatite (HA) can be successfully used to replace and augment bone tissue or as dental implants. HA is a calcium phosphate ceramic with a Ca/P ratio of 1.67 and a hexagonal structure. Mechanical properties of dense HA depend on the phase purity, density and grain size. Typically, dense HA is produced with a residual pore volume less than 5% and pore size of <1 μm. HA possesses a high hardness, moderate strengths (see Table 3.3 for compressive strength, tensile strength = 40–100 MPa) and a low fracture toughness (1 MPa.m$^{1/2}$), which limit the applications of dense HA to non-load-bearing applications and as particulates for the filling of bone defects. Due to its bioactivity (bioactivity index = 2.3), HA is also used extensively as a coating material on hip stem to promote bone fixation without the use of cement. Porous HA-based biomaterials can also be developed to imitate bone-like porous structure and are used as scaffolds in tissue engineering (bone replacement materials) [TUR 09].

3.3.4. *Resorbable ceramics*

The ultimate aim of resorbable ceramics is to degrade gradually over a period of time while being replaced by the natural tissue. In this type of material, the interfacial thickness between the material and the tissue is very thin or non-existent. When developing a resorbable device, attention has to be paid to several parameters: the resorption rates which have to be closely related to the tissue repair rates, the maintenance of the strength and the stability of the interface during the process of the material replacement by tissue, and the use of materials leading to metabolically acceptable substances during its degradation [HEN 04]. As a consequence, the composition of resorbable materials is considerably limited. Most resorbable ceramics are calcium phosphates (CaP). CaP have a very good biocompatibility due to their chemical composition containing calcium and phosphorus ions which participate in normal metabolic process. They have shown no toxicity, no fibrous tissue around the medical device and no inflammation. Resorption of CaP proceeds through various pathways: physiochemical dissolution depending on the CaP solubility and the local pH, physical disintegration into small particles, and biological degradation which induces a decrease in the local pH of the environment. Therefore, CaP that are suitable for biological implications must have a Ca/P ratio higher than 1 to possess adequate solubility and acidity, and thus appropriate hydrolysis speed. Moreover, the Ca/P ratio must be kept below 2. Indeed, tetracalcium phosphate (Ca/P = 2.0) shows too high a basicity for implantation [GRE 12].

The major CaP ceramics used in biomaterial applications are hydroxyapatite (HA), tricalcium phosphate (TCP) and a mixture of the two materials. However, stoichiometric crystalline HA has a very low solubility and no resorption occurs even after several years of implantation *in vivo*. Only resorption of poorly crystallized HA with a high specific surface area is reported. TCP possesses a Ca/P ratio of 1.5 and may crystallize in a rhombohedral structure for β-TCP or in a monoclinic structure for α-TPC. The dissolution rates of TCP are higher than that of HA, and therefore the degradation rates increase in the following order: α-TPC > β-TCP >> HA. On the other hand, TCP dissolves too fast to allow bone bonding. As a consequence, biphasic calcium phosphate ceramics composed of HA and TCP are developed. Their resorption rates are largely determined by the TCP/HA ratio with a decrease when reducing the content of TCP [THA 04]. These materials are currently available with TCP/HA ratio ranging from 1.5 to 3 with high porosity (45–70%). Due to their low mechanical properties (see Table 3.3) and high brittleness, CaP have been successfully used to replace hard tissue in low load-bearing applications, mainly as temporary bone fillers. Indeed, HA and TCP may be as resistant as cortical bone when the material porosity does not exceed 30%; but for highly porous materials (60–70%), the mechanical strength may decrease as low as 1–2 MPa. Moreover, with a porosity of 30%, TCP-based implants are still radiologically visible after 2 years of implantation, while the resorption is very fast and occurs after a few months for highly porous TCP [BON 04]. As a consequence, the surface area has to be carefully controlled to ensure appropriate resorption rates and mechanical behaviors.

3.3.5. *Glass-ionomers*

Glass-ionomers are hybrid glass polymer composites used as cements in a wide range of dental restorations since they are more esthetical than metallic materials. Glass-ionomers are constituted of fine inorganic glass particles acting as filler in an insoluble hydrogel matrix. The glass phase and the polymeric phase are held together by ionic cross-links, hydrogen bridges and entanglements of the polymeric chains [BRO 98]. The glass-ionomer cement is obtained by mixing an aqueous solution of poly(carboxylic acid), such as poly(acrylic acid), poly(maleic acid) and poly(itaconic acid), and a powder of calcium fluoroaluminosilicate glass (SiO_2-Al_2O_3-CaF_2 system). When the two components are mixed, acidic protons from the carboxylic acid groups attack the surface layer of the glass network, resulting in the release of metallic cations, mainly Al^{3+} and Ca^{2+}. Subsequently, the multivalent metallic cations neutralize the carboxylate groups of the poly(carboxylic acid) leading to the physical cross-linking of the polymeric matrix and therefore the hardening of the cement [NIC 98].

The cross-linking degree of the matrix is of major importance since it impacts the properties of the glass-ionomer cements. The reactivity of the system leans on the nature of the poly(carboxylic acid), its molecular weight and concentration, the inorganic glass/polymeric matrix ratio, as well as the glass composition. Indeed, sodium also releases from the glass and inhibits the cross-linking process by competing with calcium and aluminum in the neutralization reaction. Commercial glass-ionomers exhibit a low elastic modulus of 2–10 MPa, a compressive strength of 60–300 MPa, a flexural strength up to 50 MPa, and a fracture toughness ranging from 0.1 to 0.6 $MPa.m^{1/2}$. They have a good biocompatibility and low toxicity, and they possess anticariogenic properties since fluoride is released from the glass phase over the long term. Moreover, glass-ionomers are bioactive materials that have demonstrated a good adhesion to moist tooth structure, and they are thermally compatible with tooth enamel. Nevertheless, the main drawbacks of glass-ionomer cements are their brittleness, low modulus and wear resistance that limit their usage in stress-bearing areas [LOH 10]. Some applications of glass-ionomer cements are detailed in chapter 11.

3.3.6. *Surface processing of ceramic materials*

In contrast with metallic biomaterials, the chemical composition of ceramic surfaces is likely to be the same as that of the bulk material. However, like metals, ceramics have high surface energies and surface changes may occur by environmental contamination or corrosion. As a consequence, surface engineering processes are extensively used with ceramic materials in order to increase the hardness, wear and corrosion resistance to improve biocompatibility or to promote osseointegration [TUR 09].

As already discussed, bioinert ceramics bind to hard tissues by means of morphological fixations. To improve the bonding, the surface topography of ceramic implants can be modified to generate roughness at the macro- (10 μm–1 mm), micro- (1–10 μm) or nano- (<100 nm) scales. Since ceramics possess high hardness and chemical stability, alternative methods to the conventional ones used for topographic surface modifications of metals have to be considered. Some of these methods are mentioned below [STE 14]:

– dipping into a zirconia slurry containing a pore-former which is subsequently burned off;

– combination of sandblasting using alumina or silicon carbide particles and acid-etching treatment using combinations of HNO_3, H_2O_2 and HF;

– laser treatment which, in addition to modifying the surface topography, allows the ablation of contaminants.

Modifications of surface composition can also be used to enhance surface bioactivity. These modifications may be obtained by chemical surface treatments, such as hydroxylation or carboxylation of alumina; physical surface treatments through physical deposition, such as ion implantation in alumina, zirconia and hydroxyapatite ceramics, and CO2 irradiation of zirconia partially stabilized by magnesium (Mg-PSZ), or laser treatment. Bioactive coatings may be also applied, such as plasma-spraying of calcium phosphate, enameling using bioactive glasses or glass-ceramics, and dip-coating of sol-gel-derived Bioglass® [STE 14].

3.4. Polymers

Polymeric materials have low mechanical strength; they may deform with time and degrade. Moreover, they may contain additives, which are added to ease their manufacturing process, or low molecular weight oligomers that may lead to tissue reaction if desorbed from the material. However, polymers have unique properties in comparison with metal- and ceramic-based biomaterials. Indeed, they are resilient materials with a low density (<2.2 g.cm^{-3}), a good biocompatibility and they are easy to manufacture with various shapes [PAR 07]. Moreover, they are available in a wide variety of compositions with physical and mechanical properties that are tailored by the structure, the molecular architecture, the crystallinity degree and the transition temperature at which the material changes from a viscoelastic material to a rigid glass. Polymers represent the largest class of biomaterials and they may be synthesized or derived from natural sources [SIL 94].

Polymers may be durable or biodegradable. Non-biodegradable polymers do not undergo any chemical change *in vivo* and they are used in a wide range of biomedical applications in order to replace soft tissues, such as cartilage, vessel wall, lens, tendon and skin. Biodegradable polymers contain functional groups that may be cleaved *in vivo*, leading to the fragmentation of the polymeric chains and therefore its solubilization. Biodegradable polymers are used as temporary devices, such as resorbable sutures, bone screws and drug delivery devices, or as scaffolds in tissue engineering applications [PHU 11, PRU 11]. The use of biodegradable polymers in biomaterials applications is increasing, and the global market for tissue engineering and regeneration products is expected to reach $89.7 billion by 2016 with a CAGR of 8.4% [BCC 12]. Due to the existence of a wide range of polymers, a brief non-exhaustive description of polymeric biomaterials will be given in this chapter.

3.4.1. *Non-degradable synthetic polymers*

Non-degradable polymers are used in applications that require long-term structural stability. Some of these polymers are listed in Table 3.4 along with their mechanical and thermal properties, as well as their main applications.

Two polyolefins are widely used as biomaterials: polyethylene (PE) and polypropylene (PP). PE exists in three grades: low density, high density and ultra-high-molecular-weight PE (UHMWPE). UHMWPE is a high crystalline polymer (crystallinity >80%) with a molecular weight greater than 2×10^6 g.mol^{-1}. It is widely used as orthopedic components because of its chemical inertness, limited tissue reaction, biostability, high toughness, and good wear and fatigue resistances. Lately, cross-linked UHMWPE has been developed with a wear rate lower than that of the uncross-linked material. PP has structural properties close to PE. It is a semi-crystalline polymer with a crystallinity degree around 50–70% and a molecular weight of 2×10^5 to 7×10^5 g.mol^{-1}. It possesses a high tensile strength, a high flexural fatigue life, an excellent wear and environmental stress cracking resistances, and it is therefore used in finger joint prostheses [PAR 07]. Moreover, PP has fiber-forming characteristics which make it usable in the treatment of ventral incisional hernia [SUB 14].

Material	Chemical Structure	Tensile Strength (MPa)	Elastic Modulus (GPa)	Elongation (%)	T_g^a (°C)	T_m^b (°C)	Applications
UHMWPE	$-[CH_2-CH_2]_n-$	30-40	0.45-1.3	130-500	-100	140-150	Joint replacements
PP	$-[CH_2-CH(CH_3)]_n-$	21-40	1.0-1.6	100-300	-10	125-175	Finger joint prostheses, unabsorbable sutures
PTFE	$-[CF_2-CF_2]_n-$	15-40	0.3-0.7	250-550	-20	340	Vascular graft prostheses, heart patches
PET	$-[CH_2-CH_2-O-C(=O)-C_6H_4-C(=O)-O]_n-$	42-80	2.2-3.5	50-300	65-105	265	Vascular graft prostheses, ligaments, sutures
PEEK	$-[C_6H_4-C(=O)-C_6H_4-O-C_6H_4-O]_n-$	70-208	3.6-13	1.3-50	140	340	Suture anchors, spinal fusion
PMMA	$-[CH_2-C(CH_3)(C(=O)-O-CH_3)]_n-$	38-80	1.8-3.3	2.5-6.0	110	A^c	Bone cements, intraocular lenses, denture materials
Nylon 6/6	$-[NH-(CH_2)_6-NH-C(=O)-(CH_2)_4-C(=O)]_n-$	76	2.8	90	50	190-350	Sutures, balloons
PU	$-[C(=O)-NH-R_1-NH-C(=O)-O-R_2-C(=O)]_n-$	28-40	1.5-2.0	600-720	-80 to 140	A^c or 240	Balloons, heart valve prostheses, vascular graft prostheses, pacemaker insulation
PDMS	$-[Si(CH_3)_2-O]_n-$	6-7	<10	350-600	-50	A^c	Plastic surgery implants, ear prosthesis, facial prostheses, finger joint repairs

a Tg: glass transition temperature
b Tm : melting temperature
c A : amorphous

Table 3.4. *Main properties and applications of various non-degradable synthetic polymeric biomaterials (data from [PAR 07, SRI 09, TEO 98])*

Polytetrafluoroethylene (PTFE), known as Teflon®, is a hydrophobic crystalline polymer (crystallinity around 94% –

molecular weight of 0.5-5 × 10^6 $g.mol^{-1}$) with a low surface energy, an excellent lubricity and a high chemical inertness conferring its long-term stability. PTFE also possesses a low friction coefficient. In biomedical applications, PTFE is used in its expanded form (e-PTFE) since its microporous structure allows biointegration for fixation. Its mechanical properties vary with the porosity and the microstructure. As a consequence, e-PTFE is widely used for soft-tissue reconstruction such as vascular grafts. The main drawback of PTFE is its high density (2.2 $g.cm^{-3}$) in comparison with polyester fabrics (1.31–1.38 $g.cm^{-3}$) and its low elastic modulus which limits its use in structural components [PAR 07].

Poly(ethylene terephthalate) (PET) is a non-degradable polyester with a high melting temperature, a moderate crystallinity degree (30–40%), a high tensile strength and stiffness, and a good dimension stability. The hydrolysis of PET is restricted because of the presence of hydrophobic aromatic groups and its crystallinity. PET is mainly used in fabric or fiber forms for vascular grafts, ligaments and sutures [SUB 14].

Poly(etheretherketone) (PEEK) combines aromatic groups and flexible ether segments. It is a semi-crystalline polymer with a maximum crystallinity of 48%. PEEK possesses a good resistance to environmental stress cracking, an excellent chemical resistance, as well as high tensile and flexural strengths, a high fatigue limit, and favorable wear properties. PEEK is a thermoplastic alternative for replacing metal implant components, especially in orthopedic applications [TEO 98].

Due to their good processability and machinability, polyacrylates are used in a wide range of biomedical applications, such as intraocular lenses, bone cements for joint prosthesis fixation, denture and maxillofacial prostheses. Without any additives, poly(methyl methacrylate) (PMMA) is an amorphous polymer with a high transparency (92% transmission), a refractive index of 1.49, and a good UV light resistance, which make it appropriate in lens applications. The main advantage of PMMA as bone cement is that it can be prepared under ambient temperature. Moreover, its paste-like texture just after mixing of powdered polymer with methyl methacrylate monomer allows its penetration into the pores and open spaces of the bones. After 10 min, the cement sets and provides a secure interface between the bone and the implant [PRU 11]. However, its poor mechanical properties and wear resistance, as well as the toxicity of methyl methacrylate monomer and its exothermic polymerization, are the main drawbacks [SUB 14]. Other acrylates used in biomedical applications are poly(hydroxyethyl methacrylate) (PHEMA) and poly(octyl cyanoacrylate). PHEMA is widely used in hydrogel forms, since it can absorb water more than 30% of its weight, in intraocular lenses. Poly(cyanoacrylate) is used as a medical adhesive, replacing classical suture for cosmetic and pain reduction purposes, as well as dental cements and fillings in dentistry. Indeed, cyanoacrylate is a non-toxic resin that rapidly polymerizes in the presence of water [KAR 14].

Nylons are polyamides that have interchain hydrogen bonding. They possess a wide range of properties depending on their chemical structure and processing. They may be amorphous or semi-crystalline and have an excellent fiber-forming ability with high strength in the fiber direction. They possess excellent friction properties and good wear and abrasion resistances. However, these polymers are hygroscopic and lose their strength after implantation since water attacks the polymer amorphous phase and therefore acts as a plasticizer. As a consequence, nylons are mainly used in short-term applications such as catheter balloons in angioplasty procedure or sutures [PAR 07].

Polyurethanes (PU) are block copolymers containing low molecular weight blocks of polyethers or polyesters linked together by urethane groups. They are considered as non-degradable materials due to their poor degradation behaviors. They can be processed to obtain elastomers and glassy polymers, according to their backbone chemistry. They possess blood compatibility, high flexibility, good fatigue resistance and compliance so that they are used in short-term vascular applications [SUB 14]. PU is also widely used as coating onto metallic implants for insulating purposes in pacemaker devices [DAV 03].

Poly(dimethylsiloxane) (PDMS), known as silicone rubber, is an elastomeric polymer with a low glass transition temperature, good elasticity, flexibility and transparency, which make it appropriate for plastic surgery implants, finger joint repairs, as well as contact lenses, and wound dressing. They are used as oils, gels gums or elastomers, and the strength of the material depends on the molecular weight of the polymeric chain, and the cross-linking degree resulting from heat vulcanization [SRI 09].

3.4.2. *Synthetic and natural degradable polymers*

Biodegradable polymers are divided into two groups depending on their degradation mechanism. Most of the synthetic polymers are hydrolytically degraded since they contain hydrolytically labile chemical bonds in their backbone, such as esters, orthoesters, anhydrides, carbonates and urethanes. On the other hand, most of the natural polymers are enzymatically degraded. Biodegradable polymeric materials are used when the medical device needs to degrade over time. Various applications are considered: temporary large implants (e.g. screws, plates and contraceptive reservoirs), temporary small implants (e.g. staples, sutures and drug delivery systems) and three-dimensional porous scaffolds for tissue engineering. Due to the wide range of applications, biodegradable polymeric biomaterials have to be synthesized and designed with tailored properties in order to meet the specific requirements of each application. They must fulfill some prerequisites: do not sustain inflammatory or toxic response upon implantation, have acceptable shelf life, have degradation rate in accordance with the intended application, have appropriate initial mechanical properties with pertinent variations over the material degradation, generate non-toxic degradation products that may be metabolized and cleared from the body, and have appropriate processability and sterizability for the intended application [NAI 07].

Synthetic degradable polymers have the advantage of not presenting immunogenicity, being synthesized with more reliable sources, and being manufactured to obtain predictable properties. Some of these polymers are listed in Table 3.5 along with their mechanical and thermal properties. However, it has to be pointed out that literature review reports a wide range of mechanical properties and degradation kinetics for each individual polymer. Indeed, the material performance is affected by the polymer's molecular weight, its morphology, crystallinity, as well as the device size and shape, such as fibers, microspheres, plates and porous structures. Moreover, thermal processing and sterilization techniques also impact [DEB 08]. The main interest of natural polymers is the similarity that they share with polysaccharides or proteins composing the extracellular matrix. Collagen, chitosan, hyaluronic acid, alginate, dextran and gelatin have been widely studied for tissue engineering applications. These polymers can be recognized by the biological environment, may avoid toxicity issue, and may aid in the attachment, proliferation and differentiation of cells since they contain biofunctional molecules. However, natural polymers are complex, difficult to purify and characterize, which make them difficult to obtain as uniform raw materials. Moreover, their degradation rates are not easily controlled since enzymatic activity can vary between hosts [YOO 09].

3.4.2.1. *Degradable polyesters*

Aliphatic polyesters, such as polyglycolide (PGA), polylactide (PLA), their copolymers (PLGA) and poly(ε-caprolactone) (PCL), have been widely used in biomedical applications due to their important diversity, synthetic versatility and ease of degradation by hydrolysis of the ester linkages along the backbone. Moreover, their degradation products can be resorbed by metabolic pathways.

PGA is the simplest linear aliphatic polyester. It is highly crystalline (45–55%) with high melting and glass transition temperatures. It possesses excellent mechanical and fiber-forming properties. However, PGA degrades fairly rapidly *in vivo* (1–12 months) and leads to glycolic acid that, at high concentrations, lowers the pH of the surrounding tissue and may cause inflammation

[YOO 09]. Due to its low solubility, the use of PGA is limited to sutures and as drug delivery systems.

Material	Chemical Structure	Tensile Strength (MPa)	Elastic Modulus (GPa)	Elongation (%)	T_g[a] (°C)	T_m[b] (°C)
PGA	$\left[C(=O)-CH_2-O \right]_n$	57-920	6.5-14	1.5-25	35-45	225-233
PLA	$\left[C(=O)-CH(CH_3)-O \right]_n$	PLLA: 15-2300 PDLLA: 28-1000	PLLA: 1.2-16 PDLLA: 1.0-3.4	PLLA: 2-105 PDLLA: 2-70	PLLA: 55-65 PDLLA: 50-60	PLLA: 170-200 PDLLA: A[c]
PCL	$\left[C(=O)-(CH_2)_5-O \right]_n$	16-42	0.2-30	80 to 1000	-65 to -60	58-65
PHB	$\left[C(=O)-CH_2-CH(CH_3)-O \right]_n$	36-330	2.5-7.7	2.5-8.0	1-15	168-182
PTMC	$\left[C(=O)-O-(CH_2)_5-O \right]_n$	0.5	0.003	160	-15	A[c]
PPF	$\left[CH(CH_3)-CH_2-O-C(=O)-CH=CH-C(=O)-O \right]_n$	2-30	2-3		-20 to 32	A[c]
Polyorthoester (II)	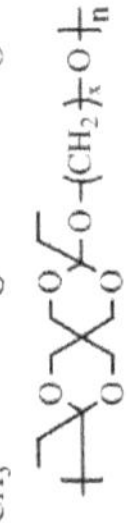	4-27	0.4-4.4	7-220	30-100	A[c]
Polyanhydride	$\left[C(=O)-R-C(=O)-O \right]_n$	4-40	0.2×10^{-3} -1.4	85	-	50-200
Polyphosphazene	$\left[N=P(R)(R) \right]_n$	2.4-7.6	0.2×10^{-3} -456		-66 to 50	242
PPDX	$\left[C(=O)-CH_2-O-(CH_2)_2-O \right]_n$	490	1.5-2.1	35	-10 to 0	106

a Tg: glass transition temperature
b Tm : melting temperature
c A : amorphous

Table 3.5. *Main properties and applications of various hydrolytically degradable polymeric biomaterials (data from [BUE 09, CHE 08, ENG 91, SRI 09, ULE 11, VAN 02, VRO 09])*

PLA may be synthesized from three isomers leading to: poly (L-lactide) (PLLA), poly(D-lactide) (PDLA) or poly(D,L-lactide) (PDLLA). PLLA is used frequently since its degradation products are similar to naturally occurring L(+) lactic acid. PLLA is a relatively hard semi-crystalline polymer (crystallinity around 37%) with a high melting temperature, while PDLLA is an amorphous polymer with only a glass transition temperature around 55°C and therefore lower tensile strength. Moreover, PLLA degrades very slowly (5 months to 5 years) and PDLLA has an intermediate degradation rate (12–16 months) between PLLA and PGA [CHE 08]. As a consequence, PLLA is generally used as screws or pins in orthopedic applications, while PDLLA is employed in drug delivery systems [PAR 07]. To modulate the mechanical properties and the degradation kinetics of PGA and PLA, different ratios of PLGA copolymers have been commercially developed and used in a wide range of biomedical applications, such as suture reinforcements, drug delivery vehicles, and skin replacement materials. PLGA with compositions between 25 and 70% GA monomer is amorphous. In these copolymers, the degradation rate decreases when the ratio of LA/GA monomers increases [NAI 07].

PCL is a semi-crystalline polyester with a low glass transition temperature that makes it semi-rigid at room temperature. PCL has low modulus and tensile strength combined with a high elongation to break, as well as a good organic solvent solubility. As a consequence, PCL may be processed to obtain various material shapes, such as microspheres, fibers and porous materials, and therefore it is used as wound closure staples, scaffolds or long-term drug delivery systems, such as one-year implantable contraceptive [ULE 11]. Indeed, PCL degrades slowly (2–3 years) in comparison with other polyesters. Its hydrolysis leads to low-concentrated caproic acid that does not cause a significant negative reaction in the surrounding tissue and that is completely metabolized since caproic acid enters the citric acid cycle. PCL also degrades through enzymatic attacks. Finally, due to its unusual properties, PCL may be compatibly blended with a wide range of other polymers and is also used as a soft block in polyurethane formulations [EDL 02].

Poly(propylene fumarate) (PPF) is a high-strength polyester containing two ester groups and one carbon–carbon double bond. The ester hydrolysis leads to fumaric acid and propylene glycol causing only mild and short inflammation. The main interest of PPF is the unsaturated double bond that can allow its covalent cross-linking. As a consequence, PPF can be used as injectable materials in tissue engineering to form scaffolds *in situ*, or as filling materials in bone defects. Moreover, cross-linked PPF possesses higher compressive and tensile strength [ULE 11]. The rigidity of the PPF chains gives a degradation of 6 months to >3 years, which depends on the cross-linking density [BUE 09].

Poly(3-hydroxybutyrate) (PHB) is a natural polyester produced biotechnologically. PHB is a high semi-crystalline polymer (crystallinity above 50%), tough and brittle. It is soluble in a wide range of solvents, and therefore it can be processed into different shapes. PHB degrades very slowly due to its high crystallinity and leads to D-(-)-3-hydroxy-butyric acid which is a normal constituent of blood. To reduce its crystallinity and result in less brittle materials with better processability, as well as to increase its degradation rate, PHB is often copolymerized with 3-hydroxyvaleric acid (HV). PHB and P(HB-HV) may find applications in long-term devices in tissue engineering of bone, cartilage, tendon, skin and nerves [ULE 11].

3.4.2.2. *Other synthetic degradable polymers*

Poly(p-dioxanone) (PPDX) belongs to the poly(ether-ester) family which is widely used as biodegradable suture materials, and also as fixation screws for small bone. PPDX is a semi-crystalline polymer (crystallinity of 37–55%) with a low glass transition temperature that makes it more flexible than PGA. The hydrolysis of PPDX occurs at the ester bond at a slow to moderately degradation rate (1–12 months) due to its high crystallinity and hydrophobicity. PPDX degradation leads to glycoxylate which is excreted in the urine or converted into carbon dioxide [NAI 07].

Aliphatic polycarbonates, such as poly(trimethylene carbonate) (PTMC), are elastomeric polymers with excellent flexibility and poor mechanical strength. The softness of the polymer makes its processing easy and allows the encapsulation of sensitive drugs under mild conditions. As a consequence, PTMC can be used in soft tissue engineering or as drug delivery systems. However, its low mechanical performance limits its applications and the polymer is generally copolymerized with other cyclic lactones. PTMC degrades slowly (>1 year) and leads to non-acidic 1, 3-propanediol and carbonic acid [EDL 02].

Polyanhydrides are the most widely investigated biodegradable polymers. Indeed, anhydride bonds are highly sensitive and polymers degrade by a surface erosion mechanism. As a consequence, polyanhydrides are used as drug delivery systems since the release of the encapsulated drugs occurs at constant rates. The chemical composition of polyanhydrides can be customized to develop materials with wide ranges of degradation kinetics. Aliphatic linear polyanhydrides degrade within a few days, while aromatic-containing polyanhydrides slowly degrade (up to 1 year). The applications of polyanhydrides are restricted due to their limited mechanical properties [HAC 08].

Poly(ortho esters) (POE) are also surface-eroding polymers used for controlled release drug delivery. They degrade very slowly since they are quite hydrophobic. POE are synthesized by the reaction of ketene acetal and an alcohol and they are divided into four classes. Direct polymerization of a triol with an orthoester leads to POE III, which are gel-like materials at room temperature due to the high flexibility of the polymeric backbone. The gel-like consistency limits the use of POE III. POE I are obtained by the transesterification between a diol and diethoxytetrahydrofuran. When degraded, POE I lead to γ-hydroxybutyric acid which autocatalyzes the polymer degradation. To overcome this issue, POE II were developed by reacting diols with diketene acetal 3,9-bis(ethylidene2,4,8,10-tetraoxaspiro[5,5]undecane). The biomaterial field is now focusing on POE IV. They are a modification of POE II through the introduction of short segments of lactic or glycolic acid into the polymer backbone. By changing the amount of acid segments, the POE IV degradation rate varies from a few days to several months [NAI 07]. These polymers are mainly limited to drug delivery applications because of their weak mechanical properties and their capacity to induce a mild-to-moderate inflammatory response [ULE 11].

Polyurethanes (PU) are a versatile class of polymers that may be non-degradable or degradable. Indeed, the biodegradation of PU depends on the chemical nature of its constituting segments. Degradable poly(ester urethanes) are developed by reacting linear diisocyanates with oligomeric diols or triols based on PLA, PGA, PCL or their combinations. The degradable PU are mainly used as micro- or nano-particulate drug delivery systems and implants for tissue engineering applications since they have significantly high degradation rates upon implantation. Composite materials may also be obtained by the addition of hydroxyapatite (HA). HA has an impact on the PU degradation behavior, as well as the differentiation of osteoblastic cells. Moreover, these composites have potential mechanical properties suitable for weight-bearing applications [SHE 14].

Polyphosphazenes are inorganic–organic hybrid polymers since they contain phosphorus and nitrogen atoms in their backbones that make the polymer hydrolytically stable, and two organic side groups attached to phosphorus that may sensitize the polymer backbone to hydrolysis. Indeed, the degradation behavior of polyphosphazenes may vary from few hours to years depending on the side group chemistry. They lead to neutral products upon degradation and have a pH buffering effect when combined with polyesters. Polyphosphazenes are flexible polymers with good processability, and they show significant promise in drug delivery for rapidly degrading polyphosphazenes, and in tissue engineering applications when substituted with hydrophobic side groups [JAM 14].

3.4.2.3. *Natural polymers*

Natural polymers derive from different sources including mammals, insects, marine organisms or plants, and may be divided into three groups: proteins, glycosaminoglycans and polysaccharides. They were used as degradable patches or sutures, and have lately generated interests as scaffolds for tissue engineering. Nevertheless, they generally degrade at fast rates from a few hours to 6 months and have low mechanical properties which limit their use to non-weight-bearing sites [BUE 09]. The chemical structures of some natural polymers are given in Figure 3.3.

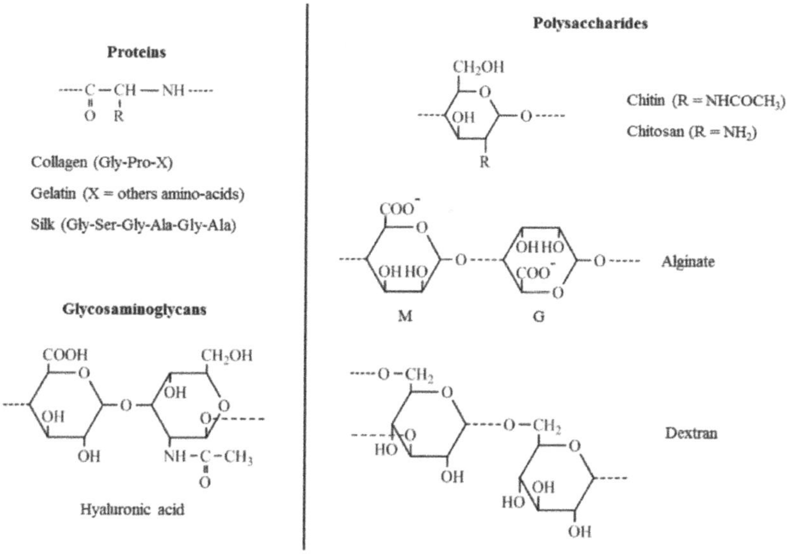

Figure 3.3. *Chemical structures of some natural polymers used in biomedical applications*

Polysaccharides have been widely used in biomedical applications since they are non-toxic, can be easily chemically modified and possess excellent properties. Chitin comes from the exoskeletons of crustaceans and is one of the most abundant natural polymers. N-deacetylation of chitin leads to chitosan (degree of deacetylation: 60–100%) whose properties depend on the source and the preparation procedure [DEB 08]. The structure of chitosan is similar to that of glycosaminoglycans present in the human body and has shown to elicit a minimal foreign-body response and to have stimulatory properties with immune cells which may stimulate wound healing process. Chitosan easily protonates and its cationic chains may be complex with various negatively charged biomolecules, which enhances its biological activity and makes it a very effective mucoadhesive [NAI 07]. Moreover, the cationic structure also confers to chitosan antimicrobial properties. Lysozyme degrades chitosan into glucosamine and the *in vivo* degradation rates range from a few weeks to 6 months depending on the degree of deacetylation [MON 11]. As a consequence, chitosan is the most promising natural polymer for tissue engineering applications such as wound dressing, and is also a possible bone graft alternative in orthopedic applications since scaffolds may be generated with predictable pore sizes and degradation rates with elastic moduli up to 11 GPa [DEB 08, BUE 09]. The main drawback of chitosan is its solubility limited to dilute acids but this property can be used to extrude viscous solutions and form gel fibers [MON 11].

Sodium alginate comes from brown algae and is a linear and non-branched block copolymer composed of variable blocks and a ratio of β-D-mannuronic acid (M) and α-L-guluronic acid (G). The polymer generally possesses a molecular weight up to 500,000 $g.mol^{-1}$ [NAI 07]. The low cost, abundance, high functionality and the easiness to form hydrogels make alginate a perfect candidate for cartilage and bone regeneration. However, the main disadvantages of alginate are the absence of bioactive sequences which may be recognized by cells, and the inability to enzymatically degrade *in vivo* since alginate lyases are found in algae and marine microorganisms [MON 11]. Moreover, alginate is too mechanically weak (elastic modulus below 27×10^{-3} GPa [BUE 09]) to be used alone for the design of scaffolds, but it may be successfully blended or copolymerized with other degradable polymers [ULE 11].

Dextran is a bacterial polysaccharide consisting essentially of α-1,6-linked glucopyranoside residues with a small percentage of α-1,3-linked residues. It has been widely used clinically as a plasma volume expander, antithrombolytic agent and drug delivery carrier. Dextran is cleaved by microbial dextranases and slowly degrades in comparison with other polysaccharides. Due to its large amount of hydroxylic groups, dextran may be easily functionalized and cross-linked. Dextran-based hydrogels are used in antifungal materials, drug delivery systems or scaffolds for tissue engineering. When designed in injectable forms, they are very promising for the delivery of biomolecules and cells. Recently, dextran found applications in nanoscience and more particularly in advanced *in vivo* imaging techniques [MAI 14].

Hyaluronic acid (HA) is the most abundant natural polymer present in the human body. It is synthesized at the inner wall of the plasma membrane and is found in the extracellular matrix, connective tissues and body fluids. HA is a linear polysaccharide composed of D-glucuronic acid and N-acetyl-D-glucosamine, and therefore it is a member of the glycosaminoglycan family although it is not sulfated. In contrast to other glycosaminoglycans, HA is not covalently bonded to proteins, and its molecular weight is high, reaching up to 10^7 g.mol^{-1}. High-molecular-weight HA is considered anti-angiogenic and non-immunogenic, while the low-molecular-weight HA is considered inflammatory, immuno-stimulatory and angiogenic. In physiological conditions, HA is negatively charged, highly hydrophilic and forms highly viscous solutions with unique viscoelastic properties. HA is insoluble in organic solvents and several modifications have been reported to alter its solubility [PED 14]. *In vivo*, HA is enzymatically degraded by hyaluronidase, β-D-glucuronidase and β-N-acetyl-hexosaminidase at a high degradation rate (from a few hours to 1 month) [MON 11, BUE 09]. HA makes itself an ideal candidate for drug delivery applications and tissue engineering, including cartilage, liver, vascular, dermal, ophthalmic and nerve repair and regeneration. Nevertheless, HA homopolymer is too mechanically weak to be used as supportive scaffolds and HA is therefore cross-linked [ULE 11].

Proteins are essentially high-molecular-weight amino acid polymers arranged in a three-dimensional folded structure. Proteins and amino acid-derived polymers have been mainly used as sutures, scaffolds for tissue engineering and drug delivery devices since they are naturally degraded by a wide range of proteases. Collagen is the most abundant protein in the human body, and its role is to maintain the structural integrity of the extracellular matrix in tissues. Its primary structure is a polypeptide chain composed of repeating triplets of Glycine-X-Y, where X and Y are typically proline and hydroxyproline. Then polypeptides arrange, via hydrogen bonding, into left-handed triple helix microfibrils which organize together in a number of different architectures to create collagen fibers. Therefore, the morphology of collagen is a 300 nm long rod with a molecular weight of 300,000 $g.mol^{-1}$ [NAI 07]. Its structural integrity and ability to withstand high tensile loads (92.5 MPa) make collagen a good candidate for tissue engineering in load-bearing applications, as well as for skin regeneration. Collagen is soluble in acidic aqueous solutions and can be processed into different forms. Moreover, due to its high reactivity, it can be cross-linked by a variety of cross-linking agents [ULE 11]. There are 28 types of collagen molecules with the most common being types I, II, III and IV. Type I represents 90% of all collagen and is found in skin, tendon and bone. Commercial sources of collagen type I are generally derived from rat tail, bovine dermis and human placenta. However, the risk of infectious diseases transmission from allogenic or xenogenic materials, the potential for immunogenicity, the high cost of purification, as well as the concern of quality and product homogeneity for mass production lead to the development of recombinant systems [MON 11].

Collagen can be converted into gelatin by denaturation and/or physical–chemical degradation. Therefore, gelatin consists of 19 amino acids and is arranged in single-stranded molecules. The process has a huge impact on the properties of the final gelatin products. Gelatin is water soluble, possesses good film forming abilities and is enzymatically degraded by collagenases since it is a derivative of collagen [VRO 09]. It has similar hemostatic properties as the collagen precursor but lacks antigenicity, and degrades much more rapidly. Gelatin is used in tissue engineering and drug delivery applications where mechanical strength is not of much importance. However, its mechanical stability may be enhanced by chemical cross-linking with various agents [MON 11].

Silk is a natural protein mainly produced by the domestic silkworm, but other natural sources may be used, such as honeybees, wasps, ants and spiders. Silk fiber consists of two types of self-assembled proteins: fibroin and sericin, containing the same 18 amino acids such as glycine, alanine and serine in different amounts. Fibroins are core proteins coated by sericin, an amorphous hydrophilic glue-like protein with a molecular weight of 300,000 g.mol^{-1}, which maintains the physical structure of fibroins. Although present in silkworm silk, sericin is absent in spider silk. The presence of sericin is a main concern since it is associated with hypersensitivity reactions and poor biocompatibility. When sericin is removed, the immune response is similar to that of other biomaterials [MON 11]. Fibroins are highly crystalline macromolecules (crystallinity around 70%) with two types of molecular weight (325,000 or 25,000 g.mol^{-1}), and containing the recurrent amino acid sequences Glycine-Serine-Glycine-Alanine-Glycine-Alanine. Fibroins are natural block copolymers constituted of hydrophobic and hydrophilic blocks which confer to silk its unique properties including high elastic modulus (5–17 GPa) and tensile strength (500–740 MPa), mechanical stability, and elongation at break around 4–20%. Silk fibroin has hemostatic properties, non-cytotoxicity, low antigenicity and non-inflammatory characteristics. Because of its crystallization and compact structure, silk degrades over a time period of several months *in vivo* and is slowly absorbed. The degradation occurs through proteolytic enzymes such as chymotrypsin, actinase and carboxylase, and therefore the rate of degradation depends on many factors, such as the silk processing conditions, the physical characteristics of the material and the implantation site. Silk can be designed in various forms and was frequently used as sutures. Recently, silk fibroin found applications in burn-wound dressings, enzyme immobilization matrices, vascular prostheses and scaffolds for tissue engineering [CAO 09].

3.4.3. *Biomedical elastomers*

Elastomers are made up of long and highly flexible polymeric chains that are cross-linked to obtain enough mechanical strength. They can withstand large deformation and recover from deformation. Their mechanical properties are tailored according to the intended application by varying different factors such as chain length, cross-linking density, nature of the polymeric chain and cross-linker. Lately, recent advances in tissue engineering have demonstrated that the behavior of cells from soft tissue is strongly influenced by the mechanical properties of the material used as substrate for cell adhesion and proliferation. Indeed, cells have greater proliferation, differentiation, as well as an encouraged cell–cell communication when the substrate possesses an elasticity that matches that of the native tissue. Moreover, when mechanical stimulation can be applied, cell behavior and the properties of the engineering tissue may be controlled. Thus, the design of biomedical elastomers has become very important since they may create a biomimetic environment favorable to cell growth and tissue development [YOU 11].

Elastomers may be cross-linked physically (thermoplastic) or chemically (thermoset). High-molecular-weight polyurethanes (PU) have been widely used as thermoplastic elastomeric biomaterials. They are synthesized from macrodiols, chain extenders and diisocyanates to develop heterogeneous multiblock structures that contain an amorphous zone providing the flexibility, while the physical cross-linking occurs through crystalline or glassy zones resulting from contributions of chain extender and diisocyanate components [YOU 11]. PU elastomers possess a wide range of chemical structures based on the chosen macrodiols which can be ether-, ester-, carbonate-based polymers, such as poly(ethylene oxide), poly(ε-caprolactone) and poly(trimethylene carbonate) [BOR 98]. PU elastomers have found applications as pacemaker wire coatings, components of artificial heart, as well as the design of scaffolds for the regeneration of cardiovascular tissue, anterior cruciate ligament and cartilage. Copoly(ether–esters), such as poly(ethylene oxide)/poly(butylene terephthalate) (PEO/PBT), are also thermoplastic elastomers. The amorphous soft segments are constituted of polyethers and the hard crystalline segments are constituted of polyesters. PEO/PBT elastomers have tensile strength ranging from 8 to 23 MPa and elongation at break from 500 to 1,300%. Recently, they found applications in dermal tissue regeneration and for cartilage engineering [YOU 11].

Thermoset elastomers are stronger materials than thermoplastic elastomers due to the covalent cross-linking. They possess homogeneous amorphous structures which confer them good mechanical stability. As described in the non-degradable polymer section 3.4.1, silicone is the most thermoset elastomers used in biomedical applications including maxillofacial prostheses, balloon catheters, finger and toe joints, pacemaker wire coatings, components of artificial heart valves, breast implants, intraocular lenses, etc. [BOR 98]. Lately, scaffolds for tissue engineering were developed by using cross-linked oligoesters. For instance, poly(ε-caprolactone)-based scaffolds can be obtained by polycondensation of polycaprolactone triol, and poly(glycerol sebacate) (PGS)-based scaffolds are cross-linked after prepolymer curing. PGS has elastomeric properties similar to cured natural rubber with a modulus of around 0.28 MPa, an ultimate tensile stress and strain of 0.5 MPa and 267%, respectively. PGS has shown promising results for the engineering of cardiovascular and nervous tissues [YOU 11].

3.4.4. *Shape–memory polymers*

The research in shape–memory polymers (SMPs) is progressing rapidly and they are very attractive to many potential applications such as sutures, clot medical devices, aneurysm occlusion devices, drug delivery systems, orthodontic therapies and smart vascular stents [SER 12]. SMPs are a special family of elastomers that are able to change their shape when exposed to a suitable stimulus. They are designed by tailored processing and programming technology to be small, rigid and easily handled during the surgical procedure at room temperature, and to become soft and elastic at body temperature and therefore to deploy into its application-relevant shape. They are of great interest since they should allow minimal invasive surgery [YOU 11]. In thermally induced SMP, the change in shape is obtained when the switching polymeric domains reach again the viscous state through the melting of the crystallites or the increase of flexibility of vitrified zones. In light-sensitive SMP, molecular switches, such as cinnamic acid, are incorporated to the polymer and form covalent cross-links. When subjected to UV light ($\lambda < 260$ nm), the cross-links are cleaved and the shape is determined by the amorphous permanent polymer. Multifunctional polymers may also be developed when additional functionalities, such as biofunctionality, hydrolytic degradability and controlled drug release, are incorporated in addition to the shape–memory effect [LEN 11].

SMPs have unique properties that make them advantageous compared to SMAs. Indeed, SMPs are lighter with a density of about 1.13–1.25 $g.cm^{-3}$, their elastic moduli reduce considerably between the glassy and rubbery states (up to 500 times), and they possess large recovery strain (up to 400%) and low recovery stresses from 1 to 10 MPa, while SMAs possess much lower recovery strain (< 8%) and higher recovery stress (1,000 MPa). Moreover, a wide field of potential applications in different thermal environments may be considered, since SMPs have a wide range of glass transition temperatures (from −70 to 100°C), as well as an excellent biocompatibility. The other major advantages of SMP are the lower cost and the easier processability by comparison with SMA. The main disadvantage of SMPs is their low recovery forces that prevent them to be used in high-power actuators [SOK 07].

Various examples of SMP systems are reported in the literature and some of them are given here. Polynorbornene was the first developed SMPs and is under consideration for ductus arteriosus occlusion and orthopedic therapy. Biodegradable SMPs may be developed to design drug delivery devices, such as ureter stents with bactericidal activity. They are generally based on diol or dimethacrylate-terminated oligomers of poly(ε-caprolactone) or poly(lactide-co-glycolide), and form a cross-linked polymer network through the use of diisocyanate or UV-curing. These systems were reported to actuate at temperature ranging from 23 to 52°C depending on the polymeric system. Devices for clot removal and aneurysm occlusion are mainly constituted of polyurethanes and actuate at higher temperature (from 32 to 85°C). Composites with nitinol were also envisioned to actuate with electro-resistive heating. Styrene-butadiene-poly(ε-caprolactone) systems actuate at 56–57°C and may undergo automatic knotting within 10 s at 70°C. Although transition temperature adjustments have to be done to consider these systems for biomedical requirements, they have high potential as thermo-sensitive sutures for wound closure [SER 12].

3.4.5. *Conjugated polymer-based biomaterials*

Conjugated polymers (CPs), also called π-polymers, are electrically and ionically conductive materials which are mechanically softer than metallic materials. Therefore, they are very attractive in the biomaterials field since they may interface ionically conducting biological tissues with electrically conductive devices, such as neural probes, cochlear implants, retinal implants and cardiac pacemakers. Indeed, these devices are mainly developed with hard metallic biomaterials that do not integrate well with soft tissues. Moreover, CPs can be more easily synthesized and processed to produce smaller devices with low electrical impedance that allows a better recording and stimulating capabilities. The other advantage of CP is the possibility to incorporate or covalently graft biological molecules that could reduce the inflammatory response and scar formation [POV 11]. Recently, CPs were investigated as transducers in biosensor applications where they mediate the electron transfer between enzymes and the final electrode, or as scaffolds for tissue engineering since cells such as fibroblasts, neurons and osteoblasts respond to electrical fields [GUI 07].

CPs may be electrochemically polymerized on the metallic surface to develop conducting CP films with a thickness of around 20 nm, or chemically polymerized to produce powders or thick films. The chemical structures of the main CP used in biomedical applications are given in Figure 3.4. The CP backbone alternates double- and single-bonded sp^2 hybridized atoms allowing the movement of electron or holes when polymers are reduced or oxidized. Prior to doping, CPs are insulative materials (10^{-10} $S.cm^{-1}$), but the type and the extent of doping can increase dramatically the electrical conductivity (Table 3.6). CPs are generally semi-crystalline polymers with mechanical properties depending on the synthesis method. Most CPs have Young's modulus and tensile strength in the range of that of thermoplastic polymers. However, their properties such as hydrophobicity, mechanical strength, malleability, roughness, redox stability and conductivity can be varied by the incorporation of dopants, the entrapment or the covalent grafting of functional groups into the polymer backbone [GUI 07].

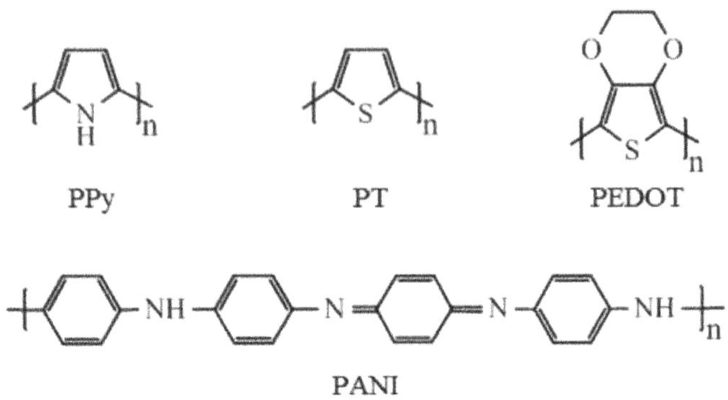

Figure 3.4. *Chemical structures of common conducting π-polymers explored for biomedical applications*

π-Polymer	Maximum conductivity ($S.cm^{-1}$)
Polypyrrole (PPy)	40–200
Polythiophene (PT)	10–100
Poly(3,4-ethylenedioxythiophene) (PEDOT)	210
Polyaniline (PANI)	5–100

Table 3.6. *Conductivity of common π-polymers explored for biomedical applications(data from [GUI 07, RAV 10])*

Polypyrrole (PPy) was extensively studied since it has high conductivity, low impedance and can be modified with additional chemical or biological functionalities. Moreover, PPy is an amorphous material. As a consequence, PPy can be used in a wide range of applications such as biosensors, drug delivery systems, bioactuators and neural prosthetics [RAV 10]. Nevertheless, PPy undergoes side reactions during electrochemical stimulation that may lead to the conjugation break of the polymer backbone [POV 11]. Poly(3,4-ethylenedioxythiophene) (PEDOT) also possesses high conductivity combined with a high-temperature stability and good functionalization possibilities. Moreover, due to the diethoxy functionalization of the thiophene ring, PEDOT is more stable and was found to be more suitable than PPy for long-term devices with needs in electrical stability. Polythiophene (PT) and polyaniline

(PANI) have lower conductivities than PPy and PEDOT. As a consequence, PT is mainly used as biosensors. PANI can be used in various applications since it is semi-flexible, may exist as bulk films or dispersions and requires simple doping chemistry [RAV 10].

3.4.6. *Polymer surfaces*

Polymeric biomaterials possess surface characteristics that may considerably differ from the bulk characteristics. Like metals and ceramics, contamination of polymeric surfaces may occur and it is generally attributed to the migration of impurities, or additives used for the polymer processing; polymeric surface may also oxidize when processed by melt techniques. Furthermore, the most striking feature is the dynamic surface of polymer materials since surface polymeric segments can reorient themselves into various conformations and rearrange their chemical groups according to the surrounding environment. Indeed, polymeric surfaces are in a thermodynamically metastable state and the conformation change occurs because polymers tend to minimize their surface energies. Surface molecules can have little up to almost liquid-like mobility, depending mainly on the polymer glass transition temperature. For instance, poly(dimethylsiloxane) possesses a very low glass transition temperature (−50°C) and a liquid-like mobility as it is in a rubbery state. In contrast, poly(methyl methacrylate), with a glass transition temperature of 110°C, has a much lower mobility since it is in a glassy state. The change in conformation is of great importance since it influences the surface chemical structure, its hydrophilicity/hydrophobicity, ionic groups, as well as the domain structure of multi-component systems. Indeed, microphase heterogeneous surfaces may arise in block copolymers. Moreover, anisotropy of polymer surfaces may be found depending on the polymer process. Finally, all these factors and the possible conformation alteration of the surface after implantation may dramatically impact the biological reactions at the interface, and as a consequence, surfaces of polymeric biomaterials need to be appropriately and intensively characterized [LYM 02].

Like metals and ceramics, many surface modifications have been developed to modify surface properties of polymeric biomaterials [DAN 14]:

– physical adsorption (Van der Waal forces, affinity binding, or electrostatic interaction): immobilizing biomolecules inducing specific cell and tissue response;

– chemical grafting (covalent grafting of poly(ethylene glycols) or polymers containing functional groups such as OH, NH_2, COOH and SH): developing antifouling surfaces, grafting of drugs, conjugating with biomolecules and creating biochemical cues;

– micropatterning (polymer demixing, block copolymer phase separation) and microtexturing (soft lithography, electrospinning, non-lithographic templating for micrometer-length scale or e-beam lithography): creating microdomains, modifying surface topography at a microscopic level, and therefore affecting cellular behavior.

Some examples of polymeric surface modifications are detailed in Chapter 6.

3.5. Conclusions

This chapter gave an overview of the principal materials used in biomedical applications. Properties, advantages and disadvantages, as well as application examples, were assessed for each material. Due to the annual growing of the global biomaterial market, research and development at the material scale are still challenging for the production of new medical devices.

3.6. Bibliography

[ASH 89] ASHBY M.F., "On the engineering properties of materials", *Acta Metallurgica*, vol. 37, no. 5, pp. 1273–1293, 1989.

[BAL 08] BALAMURUGAN A., RAJESWARI S., BALOSSIER G., *et al.*, "Corrosion aspects of metallic implants – an overview", *Materials and Corrosion*, vol. 59, no. 11, pp. 855–869, 2008.

[BAR 00] BARRAS C.D.J., MYERS K.A., "Nitinol – its use in vascular surgery and other applications", *European Journal of Vascular and Endovascular Surgery*, vol. 19, no. 6, pp. 564–569, 2000.

[BCC 12] BCC RESEARCH USA, Tissue engineering and regeneration: technologies and global markets, January 2012.

[BIE 12] BIESIEKIERSKI A., WANG J., GEPREEL M.A.-H., *et al.*, "A new look at biomedical Ti-based shape memory alloys", *Acta Biomaterialia*, vol. 8, no. 5, pp. 1661–1669, 2012.

[BOE 11] BOEHM R.D., JIN C., NARAYAN R.J., "Carbon and diamond", in DUCHEYNE P., HEALY K., HUTMACHER D.E., *et al.* (eds.), *Comprehensive Biomaterials,* 1st ed., vol. 1, Elsevier, Amsterdam, pp. 109–126, 2011.

[BON 04] BONNEVIALLE P., CLEMENT D., "Clinical applications of bone substitutes", in POITOUT D.G. (ed.), *Biomechanics and Biomaterials in Orthopedics*, Springer-Verlag, London, pp. 76–85, 2004.

[BOR 98] BORETOS J.W., BORETOS S.J., "Biomedical elastomers", in BLACK J., HASTINGS G. (eds.), *Handbook of Biomaterial Properties*, Chapman & Hall, London, pp. 302–339, 1998.

[BRE 98] BREME J., BIEHL V., "Metallic biomaterials", in BLACK J., HASTINGS G. (eds.), *Handbook of Biomaterial Properties*, Chapman & Hall, London, pp. 135–200, 1998.

[BRO 98] BROOK I.M., HATTON P.V., "Glass-ionomers: bioactive implant materials", *Biomaterials*, vol. 19, no. 6, pp. 565–571, 1998.

[BRU 04] BRUNSKI J.B., "Metals", in RATNER B.D., HOFFMAN A.S., SCHOEN F.J., *et al.* (eds.), *Biomaterials Science: An Introduction to Materials in Medicine*, 2nd ed., Academic Press, London, pp. 137–153, 2004.

[BUE 09] BUENO E.M., GLOWACKI J., "Cell-free and cell-based approaches for bone regeneration", *Nature Reviews Rheumatology*, vol. 5, no. 12, pp. 685–697, 2009.

[CAO 09] CAO Y., WANG B., "Biodegradation of silk biomaterials", *International Journal of Molecular Sciences*, vol. 10, no. 4, pp. 1514–1524, 2009.

[CAR 04] CAREW E.O., COOKE F.W., LEMONS J.E., *et al.*, "Properties of materials", in RATNER B.D., HOFFMAN A.S., SCHOEN F.J., *et al.* (eds.), *Biomaterials Science: An Introduction to Materials in Medicine*, 2nd ed., Academic Press, London, pp. 23–65, 2004.

[CAR 05] CARPENTER TECHNOLOGY CORPORATION (cartech.com), Biomaterials Alloys data – BioDur® 108 alloy, CRS Holdings Inc., 2005.

[CHE 08] CHEN Q.-Z., BRETCANU O., BOCCACCINI A.R., "Inorganic and composite bioactive scaffolds for bone tissue engineering", in CHU P.K., LIU X. (eds.), *Biomaterials Fabrication and Processing*, CRC Press, Boca-Raton, pp. 3–43, 2008.

[DAN 14] DANG T.T., NIKKHAH M., MEMIC A., *et al.*, "Polymeric biomaterials for implantable prostheses", in KUMBAR S.G., LAURENCIN C.T., DENG M. (eds.), *Natural and Synthetic Biomedical Polymers*, Elsevier, Burlington – San Diego, pp. 309–331, 2014.

[DAV 03] DAVIS J.R., *Handbook of Materials for Medical Devices*, ASM International, Materials Park, Ohio, 2003.

[DEA 07] DE AZA P.N., DE AZA A.H., PENA P., *et al.*, "Bioactive glasses and glass-ceramics", *Boletin de la Sociedad Española de Cerámica y Vidrio*, vol. 46, no. 2, pp. 45–55, 2007.

[DEB 08] DEB S., "Synthetic and natural degradable polymeric biomaterials", in CHU P.K., LIU X. (eds.), *Biomaterials Fabrication and Processing*, CRC Press, Boca Raton, pp. 3–43, 2008.

[DEM 12] DEMETRESCU I., IONITA D., PIRVU C., "Processing metallic biomaterials for a better cell response", in RAMALINGAM M., RAMAKRISHNA S., BEST S. (eds.), *Biomaterials and Stem Cells in Regenerative Medicine*, CRC Press, Boca Raton, pp. 257–280, 2012.

[DES 08] DESAI S., BIDANDA B., BARTOLO P., "Metallic and ceramic biomaterials: current and future developments", in BARTOLO P., BIDANDA B. (eds.), *Bio-Materials and Prototyping Applications in Medicine*, Springer, New York, pp. 1–14, 2008.

[DUE 99] DUERIG T., PELTON A., STOCKEL D., "An overview of nitinol medical applications", *Materials Science and Engineering: A*, vol. 273–275, pp. 149–160, 1999.

[DUM 13] DUMITRIU S., POPA V., *Polymeric Biomaterials: Structure and Function,* vol. 1, CRC Press, Boca Raton, 2013.

[EDL 02] EDLUND U., ALBERTSSON A.-C., "Degradable polymer microspheres for controlled drug delivery", *Degradable Aliphatic Polyesters, Advances in Polymer Science*, vol. 157, Springer-Verlag, Berlin-Heidelberg, pp. 67–112, 2002.

[ENG 91] ENGELBERG I., KOHN J., "Physico-mechanical properties of degradable polymers used in medical applications: a comparative study", *Biomaterials*, vol. 12, no. 3, pp. 292–304, 1991.

[GRE 12] GREMILLARD L., MEILLE S., CHEVALIER J., *et al.*, "Degradation of bioceramics", in ELIAZ N. (ed.), *Degradation of Implant Materials*, Springer, New York, pp. 195–252, 2012.

[GUI 07] GUIMARD N.K., GOMEZ N., SCHMIDT C.E., "Conducting polymers in biomedical engineering", *Progress in Polymer Science*, vol. 32, nos. 8–9, pp. 876–921, 2007.

[HAC 08] HACKER M.C., MIKOS A.G., "Synthetic polymers", in ATALA A., LANZA R., THOMSON J.A., *et al.* (eds.), *Principles of Regenerative Medicine*, Elsevier, Burlington, pp. 604–635, 2008.

[HAN 99] HANAWA T., "In vivo metallic biomaterials and surface modifications", *Materials Science and Engineering: A*, vol. 267, no. 2, pp. 260–266, 1999.

[HAU 98] HAUBOLD A.D., MORE R.B., BOKROS J.C., "Carbons", in BLACK J., HASTINGS G. (eds.), *Handbook of Biomaterial Properties*, Chapman & Hall, London, pp. 463–477, 1998.

[HEN 04] HENCH L.L., BEST S., "Ceramics, glasses, and glass-ceramics", in RATNER B.D., HOFFMAN A.S., SCHOEN F.J., *et al.* (eds.), *Biomaterials Science: An Introduction to Materials in Medicine*, 2nd ed., Academic Press, London, pp. 153–170, 2004.

[HEN 98] HENCH L.L., KOKUBO T., "Properties of bioactive glasses and glass-ceramics", in BLACK J., HASTINGS G. (eds.), *Handbook of Biomaterial Properties*, Chapman & Hall, London, pp. 355–363, 1998.

[JAM 14] JAMES R., DENG M., KUMBAR S.G., *et al.*, "Polyphosphazenes", in KUMBAR S.G., LAURENCIN C.T., DENG M. (eds.), *Natural and Synthetic Biomedical Polymers*, Elsevier, Burlington – San Diego, pp. 193–205, 2014.

[KAR 14] KARIDURAGANAVAR M.Y., KITTUR A.A., KAMBLE R.R., "Polymer synthesis and processing", in KUMBAR S.G., LAURENCIN C.T., DENG M. (eds.), *Natural and Synthetic Biomedical Polymers*, Elsevier, Burlington – San Diego, pp. 1–31, 2014.

[KNO 11] KNOWLES T.P.J., BUEHLER M.J., "Nanomechanics of functional and pathological amyloid materials", *Nature Nanotechnology*, vol. 6, no. 8, pp. 469–479, 2011.

[LEN 11] LENDLEIN A., WISCHKE C., KRATZ K., *et al.*, "Shape-memory polymers", in DUCHEYNE P., HEALY K., HUTMACHER D.E., *et al.* (eds.), *Comprehensive Biomaterials*, 1st ed., vol. 1, Elsevier, Amsterdam, pp. 479–496, 2011.

[LEV 06] LEVINE B.R., SPORER S., POGGIE R.A., *et al.*, "Experimental and clinical performance of porous tantalum in orthopedic surgery", *Biomaterials*, vol. 27, no. 27, pp. 4671–4681, 2006.

[LI 98] LI J., HASTINGS G.W., "Oxide bioceramics: inert ceramic materials in medicine and dentistry", in BLACK J., HASTINGS G. (eds.), *Handbook of Biomaterial Properties*, Chapman & Hall, London, pp. 340–355, 1998.

[LIU 04] LIU X., CHU P.K., DING C., "Surface modifications of titanium, titanium alloys, and related materials for biomedical applications", *Materials Science and Engineering: R: Reports*, vol. 47, nos. 3–4, pp. 49–121, 2004.

[LOH 10] LOHBAUER U., "Dental glass ionomer cements as permanent filling materials? -Properties, limitations and future trends", *Materials*, vol. 3, no. 1, pp. 76–96, 2010.

[LON 98] LONG M., RACK H.J., "Titatium alloys in total joint replacement – a materials science perspective", *Biomaterials*, vol. 19, no. 18, pp. 1621–1639, 1998.

[LYM 02] LYMAN D., "Characterization of biomaterials", in BARBUCCI R. (ed.), *Integrated Biomaterials Science*, Kluwer Academic/Plenum Publishers, New York, pp. 325–337, 2002.

[MAI 14] MAIA J., EVANGELISTA M.B., GIL H., *et al.*, "Dextran-based materials for biomedical applications", in GIL M.H. (ed.), *Carbohydrates Applications in Medicine*, Research Signpost, Kerala, pp. 31–53, 2014.

[MAR 05] MARQUARDT B., SHETTY R., "Beta titanium alloy processed for high strength orthopedic applications", *Journal of ASTM International*, vol. 2, no. 9, pp. 71–82, 2005.

[MAR 13] MARKETS AND MARKETS (marketsandmarkets.com), Biomaterials market – global forecasts to 2017, report BT 1556, 2013.

[MEY 08] MEYERS M.A., CHEN P.-Y., LIN A.Y.-M., *et al.*, "Biological materials: structure and mechanical properties", *Progress in Materials Science*, vol. 53, no. 1, pp. 1–206, 2008.

[MON 11] MONZACK E.L., RODRIGUEZ K.J., MCCOY C.M., *et al.*, "Natural materials in tissue engineering applications", in BURDICK J.A., MAUCK R.L. (eds.), *Biomaterials for Tissue Engineering Applications – A Review of the Past and Future Trends*, Springer-Verlag, Wien/New York, pp. 209–241, 2011.

[MOR 04] MORGAN N.B., "Medical shape memory alloy applications – the market and its products", *Materials Science and Engineering: A*, vol. 378, nos. 1–2, pp. 16–23, 2004.

[NAG 12] NAG S., BANERJEE R., "Fundamentals of medical implant materials", in NARAYAN R. (ed.), *ASM Handbook, Volume 23: Materials for Medical Devices*, ASM International, Materials Park, OH, pp. 6–17, 2012.

[NAI 07] NAIR L.S., LAURENCIN C.T., "Biodegradable polymers as biomaterials", *Progress in Polymer Science*, vol. 32, nos. 8–9, pp. 762–798, 2007.

[NIC 98] NICHOLSON J.W., "Chemistry of glass-ionomer cements: a review", *Biomaterials*, vol. 19, no. 6, pp. 485–494, 1998.

[ONG 14] ONG K.L., LOVALD S., BLACK J., *Orthopaedic Biomaterials in Research and Practice*, 2nd ed., CRC Press, New York, 2014.

[PAR 07] PARK J., LAKES R.S., *Biomaterials: An Introduction*, 3rd ed., Springer, New York, 2007.

[PAT 12] PATEL N.R., GOHIL P.P., "A review on biomaterials: scope, applications & human anatomy significance", *International Journal of Emerging Technology and Advanced Engineering*, vol. 2, no. 4, pp. 91–101, 2012.

[PED 14] PEDROSA S.S., GAMA M., "Hyaluronic acid and its application in nanomedicine", in GIL M.H. (ed.), *Carbohydrates Applications in Medicine*, Research Signpost, Kerala, pp. 55–89, 2014.

[PHU 11] PHUA K.K.L., ROBERTS E.R.H., LEONG K.W., "Degradable polymers", in DUCHEYNE P., HEALY K., HUTMACHER D.E., *et al.* (eds.), *Comprehensive Biomaterials*, 1st ed., vol. 1, Elsevier, Amsterdam, pp. 381–415, 2011.

[PIC 99] PICONI C., MACCAURO G., "Zirconia as a ceramic biomaterial", *Biomaterials*, vol. 20, no. 1, pp. 1–25, 1999.

[PIL 09] PILLIAR R.M., "Metallic biomaterials", in NARAYAN R. (ed.), *Biomedical Materials*, Springer, New York, pp. 41–81, 2009.

[POV 11] POVLICH L.K., FELDMAN K.E., SHIM B.S., *et al.*, "Electroactive polymeric biomaterials", in DUCHEYNE P., HEALY K., HUTMACHER D.E., *et al.* (eds.), *Comprehensive Biomaterials*, 1st ed., vol. 1, Elsevier, Amsterdam, pp. 547–561, 2011.

[PRU 11] PRUITT L.A., "Structural biomedical polymers (nondegradable)", in DUCHEYNE P., HEALY K., HUTMACHER D.E., *et al.* (eds.), *Comprehensive Biomaterials*, 1st ed., vol. 1, Elsevier, Amsterdam, pp. 373–379, 2011.

[RAV 10] RAVICHANDRAN R., SUNDARRAJAN S., VENUGOPAL J.R., *et al.*, "Applications of conducting polymers and their issues in biomedical engineering", *Journal of the Royal Society: Interface*, vol. 7, no. Suppl 5, pp. S559–S579, 2010.

[RIT 96] RITCHIE R.O., "Fatigue and fracture of pyrolytic carbon: a damage-tolerant approach to structural integrity and life prediction in 'ceramic' heart valve prostheses", *The Journal of Heart Valve Disease*, vol. 5, no. Suppl. I, pp. S9–S31, 1996.

[SAE 99] SAENZ A., RIVERA-MUNOZ E., BROSTOW W., *et al.*, "Ceramic biomaterials: an introduction overview", *Journal of Materials Education*, vol. 21, nos. 5–6, pp. 297–306, 1999.

[SER 12] SERRANO M.C., AMEER G.A., "Recent insights into the biomedical applications of shape-memory polymers", *Macromolecular Bioscience*, vol. 12, no. 9, pp. 1156–1171, 2012.

[SHE 14] SHELKE N.B., NAGARALE R.K., KUMBAR S.G., "Polyurethanes", in KUMBAR S.G., LAURENCIN C.T., DENG M. (eds.), *Natural and Synthetic Biomedical Polymers*, Elsevier, Burlington – San Diego, pp. 123–144, 2014.

[SIL 94] SILVER F.H., *Biomaterials, Medical Devices & Tissue Engineering: An Integrated Approach*, Chapter 1, Chapman & Hall, New York, 1994.

[SOK 07] SOKOLOWSKI W., METCALFE A., HAYASHI S., *et al.*, "Medical applications of shape memory polymers", *Biomedical Materials*, vol. 2, no. 1, pp. S23–S27, 2007.

[SRI 09] SRICHANA T., DOMB A.J., "Polymeric biomaterials", in NARAYAN R. (ed.), *Biomedical Materials*, Springer, New York, pp. 83–119, 2009.

[STE 14] STEFANIC M., KOSMAČ T., "Surface modifications of load-bearing ceramics for improved osseointegration", in SHEN J.Z., KOSMAČ T. (eds.), *Advanced Ceramics for Dentistry*, Elsevier - Butterwort-Heinemann, Waltham, pp. 301–325, 2014.

[SUB 14] SUBRAMANIAM A., SETHURAMAN S., "Biomedical applications of nondegradable polymers", in KUMBAR S.G., LAURENCIN C.T., DENG M. (eds.), *Natural and Synthetic Biomedical Polymers*, Elsevier, Burlington, San Diego, pp. 301–308, 2014.

[TEO 98] TEOH S.H., TANG Z.G., HASTINGS G.W., "Thermoplastic polymers in biomedical applications: structures, properties and processing", in BLACK J., HASTINGS G. (eds.), *Handbook of Biomaterial Properties*, Chapman & Hall, London, pp. 270–301, 1998.

[THA 04] THAMARAISELVI T.V., RAJESWARI S., "Biological evaluation of bioceramic materials–a review", *Trends in Biomaterials and Artificial Organs*, vol. 18, no. 1, pp. 9–17, 2004.

[TUR 09] TURNER I.G., "Ceramics and glasses", in NARAYAN R. (ed.), *Biomedical Materials*, Springer, New York, pp. 3–39, 2009.

[ULE 11] ULERY B.D., NAIR L.S., LAURENCIN C.T., "Biomedical applications of biodegradable polymers", *Journal of Polymer Science Part B: Polymer Physics*, vol. 49, no. 12, pp. 832–864, 2011.

[VAN 02] VAN DE VELDE K., KIEKENS P., "Biopolymers: overview of several properties and consequences on their applications", *Polymer Testing*, vol. 21, no. 4, pp. 433–442, 2002.

[VRO 09] VROMAN I., TIGHZERT L., "Biodegradable polymers", *Materials*, vol. 2, no. 2, pp. 307–344, 2009.

[WAN 04] WANG M., "Bioactive materials and processing", in SHI D.L. (ed.), *Biomaterials and Tissue Engineering*, Springer-Verlag, Berlin, 2004.

[YAN 10] YANG K., REN Y., "Nickel-free austenitic stainless steels for medical applications", *Science and Technology of Advanced Materials*, vol. 11, no. 1, p. 014105 (13pp), 2010.

[YOO 09] YOON D.M., FISHER J.P., "Natural and synthetic polymeric scaffolds", in NARAYAN R. (ed.), *Biomedical Materials*, Springer, New York, pp. 415–441, 2009.

[YOU 11] YOU Z., WANG Y., "Bioelastomers in tissue engineering", in BURDICK J.A., MAUCK R.L. (eds.), *Biomaterials for Tissue Engineering Applications – A Review of the Past and Future Trends*, Springer-Verlag, Wien–New York, pp. 75–118, 2011.

4

Biocompatibility and Norms

4.1. Introduction

Going back to the first and second generations of biomaterials, which appeared before the birth and impressive development of the field of biomaterials, biomaterials were the materials used to replace a part or all of a deficient organ, and were chosen because they were easily available and exhibited the right or at least the best mechanical properties and resistance to corrosion for one application. This is the case of metals such as stainless steel and polymers such as poly(tetrafluoroethylene) and poly(methyl methacrylate) used in dentistry and orthopedics, of polymer fabrics or knittings such as poly(ethylene terephthalate) used in vascular surgery from the 1950s, of polymers such as poly(methyl methacrylate) used in ophthalmology, etc. All these materials shared the same characteristic: they had not been specifically engineered for biomedical applications.

Since the second part of the 20th Century, the techniques of synthesis and elaboration of materials have helped in proposing new biomaterials with improved properties, due to the better mastery of chemical and physical processes. At the same time, the interest in other properties emerged regarding the whole behavior of the material when implanted or placed in the living system. This was emphasized

---

Chapter written by Véronique MIGONNEY.

by the postoperative reactions observed by the pioneering surgeons in the field of biomaterials. They studied and reported on the influence of the material reaction on the living system functions such as the coagulation system, the inflammatory response, the immunity defense system, etc. Secondary effects having no desired and even dangerous consequences on the health of the patient were identified and differentiated: the effects of the biomaterial on the host – leakage of sub-products or simple contact leading to inflammation, allergy, necrosis of the tissues, generation of wear debris responsible for chronicity of the inflammatory response, general toxicity of different products or materials and infection – and the effects of the host environment on the biomaterial – degradation, corrosion, wear, cracking, solubilization, aging, etc.

Then, biomaterials could be "classified" as a particular category of materials needing specific properties for different applications in medicine. All the studies, information and the amount of experience from scientists and medical personnel showed that serious controls of the material (quality, purity, toxicity) and the perfect understanding of the consequences of an implantation were needed to prevent failure. This led to the building of the concept of "biocompatibility", which represents the required properties of a biomaterial for its use in the biological environment: the elaboration of a biomaterial cannot be disconnected from the notion of biocompatibility. In the same manner, any biomaterial has to be conceived bearing in mind its sterilization process.

4.2. Definitions of "biocompatibility"

4.2.1. *Introduction to biocompatibility*

Biocompatibility is composed of the required properties that any biomaterial has to exhibit when implanted in the living tissues of a human body. It represents the ability of the material to coexist in contact with the tissues without causing deleterious effects that could compromise the health and function of the tissues. Nevertheless, depending on the use of biomaterials – short or long term, invading or not, location of the implantation – the requirements can range from low and acceptable to very drastic (as soon as blood circulation and contact are involved) because the consequences may be low or significantly dramatic.

To help researchers, surgeons, industrial partners, medical personnel and patients respond to this question and understand the concept of biocompatibility, those who defined biomaterials also worked on the concept of biocompatibility since a biomaterial does not present any interest if it is not associated with its degree of biocompatibility.

4.2.2. *History of the definitions*

Biocompatibility should be considered as a set of properties that are necessary and required if one material is destined to be used as a biomaterial. There is a constant improvement in the definitions of biocompatibility: the earliest definitions described no desired effects of the biomaterial with the living system, which was basically not true of the real property of biocompatibility.

For many years, biocompatibility was defined by default as the list of the effects which must not be engendered by the presence of a biomaterial or ignored. It is often associated with:

– the release or leakage of products or by-products of degradation or corrosion, or additives, or contaminants;

– the biological activity and the further consequences of these releases locally or not.

Then, materials to elaborate biomaterials were selected or developed on the basis of their non-toxicity, non-thrombogenicity, non-carcinogenicity, non-irritability, etc.

The question that arises is: "Does only one definition of biocompatibility exist?" Several parameters introduced the re-evaluation of the definition of biocompatibility:

– it appeared obvious that the biological response to specific materials can vary from one application to another and from one implantation site to another. Therefore, biocompatibility cannot exclusively depend on the characteristics of the material and must be defined according to the application to which the biomaterial is intended;

– in the majority of cases, the material has to specifically respond and interact with surrounding tissues rather than be inactive or passive or inert;

– in a similar context, some particular applications require a degradation of the material over time rather than staying infinitely biodegradable biomaterials.

With the expanding experience in biomaterials science, the definition has evolved to that proposed in 1987 by D. Williams [WIL 87]: “Biocompatibility refers to the ability of a material to perform with an appropriate host response in a specific application”.

This is an exact definition of biocompatibility’s point of view Nevertheless, biomaterials are not exclusive to the research domain and many people use and abuse the word “biocompatibility” or the adjective “biocompatible”. Sometimes, it does not match reality. Why? Is biocompatibility possible for all implanted materials? What is an appropriate host response?

4.3. Discussion on biocompatibility

Why can biomaterials not be dissociated from the notion of biocompatibility? The reason is that when introduced in the body of a human being, biomaterials inevitably lead to interactions with the biological systems, the other name of which is “the host response”. The host response (see below) is a very complex set of reactions.

For a long time, biomaterials were classified according to their reactivity toward living systems. A biomaterial could be qualified as:

– bioinert: if its role was restricted to the replacement of one or several functions of an organ, such as cardiac valves;

– bioactive: if the biomaterial was able to develop strong interactions with the surrounding environment of implantation – hydroxyapatite (HAP) coating on hip prostheses (HAP is present in mineralized bone) improves the biocompatibility and integration of the implant in the osseous tissues. Bioactivity exists and has been shown in several applications!

One of the above classifications is irrational because the bioinertness of a material place in living tissue cannot exist. The reason is clear: as soon as a material is placed in contact with a biological environment, it induces a host response – good or bad, appropriate or hostile. Bioinertness does not exist.

Discussion on biocompatibility: what is the goal of the biocompatibility assessment? What is it moving towards? The above definitions are not so precise. The measurement of biocompatibility in a laboratory is not well defined as normalized. To understand and measure the biocompatibility of materials is a particular domain of biomaterials science.

To understand that point of view, it is necessary to invest in the interdisciplinary biomaterials science that is at the interface of several disciplines such as materials science, biology and medicine. If it is impossible to have knowledge of all of those disciplines, some notions are nevertheless essential. The study of the host response helps in differentiating the quality and properties of biomaterials.

4.4. Host response

The definition of a host is "a person who receives guests in his own home"; in biology, host means "an animal or plant that nourishes and supports a parasite". In the context of biomaterials, the "host response" is all the responses of the body of a human being following the implantation of a biomaterial [AND 84, AND 08].

The host response to biomaterials has been perfectly described by Ratner *et al.* [RAT 04] and Anderson *et al.* [AND 08].

The inflammatory response is the natural defense response of the body to any injury or to attack by a foreign body such as bacteria or viruses or even biomaterials. The different steps of this response and the consecutive healing process are well defined:

(1) the injury (wound, surgical procedure);

(2) acute and chronic inflammation (step one of the defense);

(3) foreign body reaction (step two of the defense = the wound healing begins);

(4) granulation of the tissues (step three of the defense = new connective and vascularized tissues are formed);

(5) fibrosis and encapsulation (step four of the defense = a fibrous barrier is created between the foreign body and the rest of the living system, this is a hostile host response).

The role of all these well time-scheduled steps is to fight against and/or destroy foreign bodies. From the point of view of biomaterials, the wound can be caused by the introduction of an implant in the body, by injection or insertion. Then, macrophages that naturally play a major role for phagocyte bacteria or viruses or undesired species cannot, in this particular case, go on to phagocytosis.

The defense system is disrupted by the presence of the biomaterial implant and the host response is then no longer controlled. The inflammation goes from acute to chronic: naturally activated macrophages go on being activated in the tissues surrounding the implant, cytokines are constantly secreted, macrophages form giant cells intending to phagocyte the oversized foreign body which is the implant. Macrophages, giant cells and cytokines coexist and induce the recruitment of fibroblasts, the role of which being to secrete collagen and encapsulate the biomaterial.

The biomaterial is progressively surrounded and encapsulated by fibrous tissues or a fibrous barrier and is kept isolated from the normal and differentiated tissues. Fibrosis and encapsulation represent the final stage of the inflammatory response to biomaterials. It is characterized by a cell adaptation, a phenotypic change as well as an overproduction of proteins such as collagen and ECM. Fibrosis works as a repair and restoration of a structure but cannot reproduce the initial function of the tissue. The biomaterial is not well integrated into the tissues – it is not "biointegrated" – it cannot be qualified as biocompatible since a hostile host response or foreign body response has been developed against the biomaterial. Is the biomaterial biotolerant or biocompatible? Does an encapsulation mean a disrupted healing or an abnormal healing process? The intensity of the inflammatory host response gives an idea of the biocompatibility of a material.

Aside from the inflammatory and immune system, the other part of the host response is more specific and depends on the location of the implantation. Coagulation is an important part of the host response since it facilitates the regulation and equilibrium of the entire blood vascular system. In the case of contact between a biomaterial and blood, a cascade of reactions lead to the formation of a blood clot – biocompatibility with blood, or "hemocompatibility", would mean the biomaterial helps in preventing the formation of a thrombus. This biomaterial does not yet exist even if several surface modifications have been set up to solve this problem.

As the host response is developed at the interface between the living tissues and the implant, it is strongly dependent on the nature of the biomaterial surface: chemistry, physico-chemistry (hydrophilic/hydrophobic balance), structure (rough, smooth, nano- or microstructuration). The most extreme care has to be taken on the control of the surface.

4.5. Biocompatibility – how can we evaluate it?

Biocompatibility is very often correlated with performance and success in a particular domain or a precise application [WIL 08, PAR 05]. Therefore, to *evaluate biocompatibility, it is useful to keep in mind the definition of biocompatibility given by D. Williams* in section 4.2.2. That definition insists on the specificity of the application to which the biomaterial is intended. Although not explicit, this definition notes the duration of the implantation.

Since biomaterials and their applications are numerous, the evaluation of their biocompatibility may include different degrees of requirements [ZIA 88].

Biocompatibility is multifactorial and has to take material, tissue, chemical composition, nature of the surface and mechanical constraints into account. A material implant that is considered biocompatible for one application can be totally inappropriate for another.

How can we determine if a biomaterial is biocompatible and if it correctly performs its functions in the *in vivo* environment? What duration: short- or long-term implantation? Which application in surgery: cardio-vascular, orthopedic, ophthalmologic, cosmetic, urologic, dental? Which animal model? The questions are of the utmost importance and must be asked, studied and solved from the beginning of the study and this will lead us to set up a protocol of evaluation of biomaterials. Various procedures of assessment and testing exist because some biomaterials are intended to be implanted for very short terms and have to fill their functions only for seconds, while others will be implanted for life and have to maintain their performance for the longest possible time.

Another point to be studied is the evaluation of the importance of the successes of the *in vitro* results. What must all *in vitro* experiments include? Some tests could be ignored or the test may not exist. When and how do we decide to undertake much heavier and more expensive *in vivo* evaluation? For the *in vivo* evaluation of a biomaterial implant, it is difficult to make the choice of an animal and the duration of the implantation. The cost of these tests and the ethical rules and protocols are necessary to optimize the time of the implantation and the nature of the animal as a function of the further application and use. Mice, rats and rabbits are useful animal models; nevertheless, these models could be far from the future conditions of use of the biomaterial such as in term of its mechanical performances and its reliability. Along the same idea, an *in vivo* evaluation normally has to be useful in terms of results on the behavior in time, whereas it is neither easy nor possible to undertake *in vivo* experiments too long; preclinical studies, when possible, are the most reliable experiments that can help in giving the response to the future success or failure of long-term implantations.

The evaluation of biocompatibility can be done in research laboratories in the framework of basic research programs on one material for one application. In this case, the biocompatibility assessment is generally carried *in vitro* and *in vivo* and leads to the development of an implant, the properties and characterization of which are very precise and controlled. *In vitro* and *in vivo* tests give a good idea of the performances and safety of the biomaterial (mechanical test, biocompatibility, aging, sterilization). This is possible when the whole study is done in constant collaboration between chemists, physicians, veterinarians, surgeons and industrial partners, and it allows the development of some animal models that have not yet been described in the ISO NORMS (see Chapter 6 on animal models). One question about the animal experiments concerns the duration of the animal experiment and the correlation of the results to the future implant duration – temporary, short or long term.

4.6. Infection, sterilization, prevention of infection

As intended to be placed in the body of a human being, the biomaterial implant has to be sterile to prevent further infection, disease and patient death. Therefore, a biomaterial must be rapidly considered in the context of its final fabricated and sterilized form.

There are several validated processes of sterilization being used every day in medical environments such as hospitals, surgeries to guarantee the safety of medical tools. These processes have been adapted to the case of biomaterials: the process of sterilization will depend on the chemical nature of the biomaterial device to prevent early degradation of the material and release of degradation products and compounds in the body [LER 12].

Then, among the different sterilization processes, the choice will be made according to the physico-chemical characteristics of the devices as well as the facilities where the sterilization will be carried out – research laboratory, surgery, industrial plant, hospital. All these parameters have to be pointed out because the advantages and disadvantages of sterilization can affect the in fine "biocompatibility" of the material and any biomaterials have to be conceived bearing in mind its sterilization process.

Some major questions have to be answered and taken into account:

– sterilization process and results as part of the biocompatibility evaluation of a biomaterial;

– methods of sterilization/the right choice for each biomaterial: does sterilization process change the "initial" biocompatibility of a biomaterial?

– degree of risk.

Infectious risk is dependent on the nature of the tissues with which the biomaterial will be in contact. There are three levels of risk to be taken into consideration and three levels of requirements for the sterilization treatment (see Table 4.1).

Condition of use	Risk
Contact with blood or vascular system or vascularized tissues	High risk
Contact with mucous membranes or a skin wound	Medium risk
Contact with healthy skin	Low risk

Table 4.1. *Infection risk level of a medical device according to its use*

Implanted biomaterials are considered as high risk devices in terms of infection. Extensive washings and cleanings have to precede the sterilization process. A biomaterial which has not been well washed and cleaned cannot be sterilized with safety.

Cleaning and decontamination are different from disinfection and sterilization:

– decontamination: this means a bacteriostatic action;

– disinfection: this means a bactericidal action. It is obtained by using disinfectant products – the effect of which being the destruction of the dangerous germs – the germicide action has to respond to the standards (ex: NF T AFNOR 72);

– sterilization: this is an operation that allows us to eliminate or kill microorganisms so that a product or an object becomes sterile. A medical device or a biomaterial can be labeled sterile if it is exempt from any microorganism and the theoretical probability that a viable microorganism is present on this device must be equal to or lower than $1/10^6$.

The sterilization is a complex multi-step process which gives a sterile product and keeps it in a sterile state for a specified period of time (e.g. Norm AFNOR NF T 72-101).

Sterilization is one of the mandatory processes to keep bacteria and infection away from biomaterial implants; the sterilization process has to be validated (see NORMS). Nevertheless, despite the very strict disease prevention methods to fight biomaterial implant infections – washings, sterilization and antibiotics – a percentage of implants are still infected. The infection when installed may turn out to be very difficult to treat, leading to complex antibiotics mixture treatments, and in extreme cases the removal of the implant is then necessary to eradicate the infection. Bacterial infection is one of the major problems encountered after the implantation of biomaterials such as catheters, devices of extra corporeal recirculation, renal dialysis systems, contact lenses, intraocular lenses, joint prostheses. Catheters infections are one of first causes of nosocomial bacteraemia and are responsible for a significant mortality.

For several years, researchers have worked on the functionalization of biomaterial surfaces in order to prevent infection [BER 02, MIG 06, CRE 03, PAV 01, PIC 12]. The advantages of these prevention methods are numerous but one is worthy of note: it may prevent the use of antibiotics and thus help in the fight against bacterial resistance to antibiotics, which is a major problem for public health.

4.7. Norms and biocompatibility?

As biomaterials belong to the medical device family, they have to follow the requirements of the different classes of medical devices described in ISO NORMS EN ISO 10993. The classification dividing the medical devices in class I to III as a function of the risk encountered by the patient and the ISO NORMS gives the guidelines to establish the assessments to be done on one biomaterial depending on its application: the *in vitro* and *in vivo* experiments are precisely described for each class of biomaterial as a function of its application. These norms and the norms developed in other countries such as European countries respond to the necessity of guaranteeing the quality of the biomaterials that are proposed to the surgeons for short- or long-term implantations. Each new material, each new functionalization process or each change in the composition of a material or a biomaterial needs a new evaluation of its performance when placed in contact with a biological and living system.

The European EN ISO 10993 Norms group 20 parts (see Table 4.2) which specify and detail the different series of standard methods of evaluation of biocompatibility a medical device has to pass prior to clinical studies; in the case of passing all the tests, the medical device can be CE marked. The CE marking provides an indication that the assessments have been carried out, before the product is placed on the market. It must be affixed visibly to the product. All these rules make it possible to ensure its compliance with the legislative requirements.

Norm number	*Date/Up dated/ amendment*	*Part*	*Object*	*Tests*
ISO 10993-1	2009	1	Biological evaluation of medical devices	Evaluation and testing in the risk management process
ISO 10993-2	2006	2	Biological evaluation of medical devices	Animal welfare requirements
ISO 10993-3	2003	3	Biological evaluation of medical devices	Tests for genotoxicity, carcinogenicity and reproductive toxicity
ISO 10993-4	2002 Amd 1:2006	4	Biological evaluation of medical devices	Selection of tests for interactions with blood
ISO 10993-5	2009	5	Biological evaluation of medical devices	Tests for *in vitro* cytotoxicity
ISO 10993-6	2007	6	Biological evaluation of medical devices	Tests for local effects after implantation
ISO 10993-7	2008		Biological evaluation of medical devices	Ethylene oxide sterilization residuals
ISO 10993-8	2001	8	Biological evaluation of medical devices	Selection of reference materials
ISO 10993-9	1999	9	Biological evaluation of medical devices	Framework for identification and quantification of potential degradation products
ISO 10993-10	10:2010 Part 10:		Biological evaluation of medical devices	Tests for irritation and delayed type hypersensitivity
ISO 10993-11	2006	11	Biological evaluation of medical devices	Tests for systemic toxicity
ISO 10993-12	2012	12	Biological evaluation of medical devices	Sample preparation and reference materials (available in English only)

ISO 10993-13	1998	13	Biological evaluation of medical devices	Identification and quantification of degradation products from polymeric medical devices
ISO 10993-14	2001	14	Biological evaluation of medical devices	Identification and quantification of degradation products from ceramics
ISO 10993-15	15:2000	15	Biological evaluation of medical devices	Identification and quantification of degradation products from metals and alloys
ISO 10993-16	1997	16	Biological evaluation of medical devices	Toxicokinetic study design for degradation products and leachable
ISO 10993-17	2002	17	Biological evaluation of medical devices	Establishment of allowable limits for leachable substances
ISO 10993-18	2005	18	Biological evaluation of medical devices	Chemical characterization of materials
ISO/TS 10993-19	2006	19	Biological evaluation of medical devices	characterization of materials
ISO/TS 10993-20	2006	20	Biological evaluation of medical devices	Principles and methods for immunotoxicology testing of medical devices

Table 4.2. *ISO Norms 10993 – biological evaluation of medical devices*

Biomaterials may be distributed in four classes according to the risk they pose to the patient: Class I, Class IIa, Class IIb and Class III (article R5211-7 of the French Public Health Code). The majority of, implanted biomaterials belong to Classes II to III:

– Class IIa: medium risk – invasive devices – example: contact lens, dental crown;

– Class IIb: high risk – invasive devices – long-term implantation – examples: internal sutures;

– Class III: very high risk – invasive devices – contact with blood, vascular or nervous central system, devices incorporating medicine, long-term implant or biodegradable implant, device incorporating animal derivative tissues – example: stent, hip prosthesis, ligament, joint prosthesis and so on.

4.8. Conclusion

Biomaterials lead to various and numerous biological reactions and interactions that have to be controlled to be qualified as biocompatible.

"Biocompatibility" groups a set of performances that a biomaterial has to assure to be used in therapeutic purposes and to improve the quality and duration of the life of a patient.

In research laboratories, "biocompatibility" can be evaluated through "specific" *in vitro* and *in vivo* tests to conclude on a mechanism of the observed biocompatibility. The initial product is a laboratory "product", the biological properties of which are to be evaluated. What tests have to be carried out? *In vitro* tests or *in vivo* tests? How can we realize them? In addition to the biological properties, some additional studies and analyses regarding the mechanical properties, the stability of the physico-chemical properties under physiologic conditions, the effect of the sterilization on its physical, chemical, biological properties have to be performed. When the laboratory "product" is successful, what continuation is to be given? Clinical trial? CE or FDA? When the biomaterials are intended to be implanted in a patient, the biocompatibility has to be validated and this is done by the series of tests imposed by Norms – ISO EN NORMS 10993 in Europe.

Despite all the proposed and imposed experiments and controls, failures may be and are encountered. Among the causes of failure:

– some may be directly attributed to the biomaterial or be a consequence of the implantation of the biomaterial;

– some may put the patient's life rapidly in danger, such as thrombosis, thromboembolism, bleeding and infection;

– some do not put the life of the patient in danger, such as fibrosis and encapsulation, aseptic loosening (inappropriate repairs), degradation of the material (wear debris, rupture, tear), migration of the implant;

– some are hostile host responses, such as inflammation, toxicity, teratogenicity, tumorigenic, dedifferentiation of the surrounding tissues (calcification), allergy.

The failures translate the inadequacy of the biomaterial properties and characteristics of the function of implant. This is the reason why a strong importance has to be put upon:

– the stability of the material in the physiological environment;

– the constancy of the characteristics of the material – elaboration/fabrication, conditioning, packaging, sterilization, storage and so on;

– the animal models and preclinical tests;

– the implantation and surgical techniques.

These above examples show to what extent the concept of biomaterials and biocompatibility are complex. Some biomaterials were successfully used for one or several applications such as poly(methyl methacrylate) used in 1937 in dentistry then in ophthalmology, poly(ethylene terephthalate) used since 1958 in the elaboration of vascular prostheses and then in ligaments and poly(ethylene) used since 1960 as a part of hip prostheses.

4.9. Bibliography

[AND 84]) ANDERSON J.M., MILLER K.M., "Biomaterial biocompatibility and the macrophage", *Biomaterials*, vol. 5, no. 1, pp. 5–10, January 1984.

[AND 08] ANDERSON J.M., RODRIGUEZ A., CHANG D.T., "Foreign body reaction to biomaterials", *Semin Immunol.*, vol. 20, no. 2, pp. 86–100, April 2008.

[BER 02] BERLOT S., PAVON G., HELARY G., *et al.*, "Biomimetic PMMA based terpolymers: modulation of bacterial adhesion effect chemical composition and monomers distribution effect", *Biomacromolecules*, vol. 3, no. 1, pp. 63–68, 2002.

[CRÉ 03] CRÉMIEUX A.-C., PAVON-DJAVID G., SALEH MGHIR A., *et al.*, "Bioactive polymers grafted on silicone to prevent Staphylococcus aureus prosthesis adherence: in vitro and in vivo study", *Journal of Applied Biomaterials & Biomechanics*, vol. 1, pp. 178–185, 2003.

[EVA 02] EVANS M.D., MCFARLAND G.A., XIE R.Z., *et al.*, "The use of corneal organ culture in biocompatibility studies", *Biomaterials*, vol. 23, no. 5, 1359–1367, March 2002.

[GOO 08] GOODMAN S., "Wear particulates and osteolysis?", *Orthop Clin North Am*, vol. 36, no. 1, pp. 41–48, 2008.

[LER 12] LEROUGE S., "Introduction to sterilization: definitions and challenges", in LEROUGE S., SIMMONS A. (eds.), *Sterilization of Biomaterials and Medical Devices*, Woodhead Publishing, 2012.

[MIG 03] MIGONNEY V., PAVON G., SIOVE A., *et al.*, Procédé de traitement de surfaces polymères, French Patent No. 0300495, 2003.

[MIG 06] MIGONNEY V., NOIRCLÈRE F., HÉLARY G, Procédé de greffage de polymères bioactifs sur des matériaux prothétiques, French Patent No. FR 06 05 073, June 2006.

[MIG 09] MIGONNEY V., HELARY G., "Bioactive polymer coating to improve bone repair", in PLANELL J.A., BEST S., LACROIX D., *et al.* (eds.), *Bone Repair Biomaterials*, Woodhead Publishing, pp. 309–323, 2009.

[PAR 05] PARIENTE J.L., BORDENAVE L., CONORT P., "Biomateriaux et génie tissulaire en urologie – Partie A: Aspects généraux – Chapitre III: Biomatériaux, Biomatériels", *Biocompatibilité, Prog Urol*, vol. 15, no. 5, pp. 887–890, 2005.

[PAV 01] PAVON G., HELARY G., MIGONNEY V., *Biomatériaux modifiés: utilisation thérapeutique actuelle et future, Infections à Staphylocoques sur prothèse*, EDK Medical and Scientific Publishers, p. 41, 2001.

[PIC 12] PICHAVANT L., AMADOR G., JACQUELINE C., *et al.*, "pH-controlled delivery of gentamicin sulfate from orthopedic devices preventing nosocomial infections", *J Control Release*, vol. 162, no. 2, pp. 373–381, September 2012.

[RAT 04] RATNER B., HOFFMAN A, SCHOEN F., *et al.*, *Biomaterials Science: An Introduction to Materials in Medicine*, 2nd ed., Elsevier Academic Press, 2004.

[RAT 07] RATNER B.D., "The catastrophe revisited: blood compatibility in the 21st Century", *Biomaterials*, vol. 28, no. 34, pp. 5144–5147, December 2007.

[RAT 11] RATNER B.D., "The biocompatibility manifesto: biocompatibility for the twenty-first century", *J Cardiovasc Transl Res.*, vol. 4, no. 5, pp. 523–527, October 2011.

[WIL 87] WILLIAMS D., *Definitions in Biomaterials*, Elsevier, Amsterdam, 1987.

[WIL 99] WILLIAMS D., *The Williams Dictionary of Biomaterials*, Liverpool University Press, Liverpool, 1999.

[WIL 08] WILLIAMS D., "On the mechanism of biocompatibility", *Biomaterials*, vol. 29, pp. 2941–2953, 2008.

[VER 12] VERT M., DOI Y., HELLWICH K.H., *et al.*, "Terminology for biorelated polymers and applications (IUPAC recommendations 2012)", *Pure Appl. Chem.*, vol. 84, no. 2, pp. 377–410, 2012.

[ZIA 88] ZIATS N.P., MILLER K.M., ANDERSON J.M., "In vitro and in vivo interactions of cells with biomaterials", *Biomaterials*, vol. 9, no. 1, pp. 5–13, January 1988.

[ZHA 08] ZHANG Z., ZHANG M., CHEN S., *et al.*, "Silk-functionalized titanium surfaces for enhancing osteoblast functions and reducing bacterial adhesion", *Biomaterials*, vol. 29, no. 36, pp. 4751–9, December 2008.

5

Bioactive Polymers and Surfaces: A Solution for Implant Devices

5.1. Introduction

The work on bioactive implant devices reported here has been developed for around 30 years with the aim of "ensuring" a control of the *in vitro* and *in vivo* biological response by modifying the surface chemistry of macromolecules or prosthetic surfaces through macromolecular synthesis or grafting of *"bioactive" polymers* on polymer or metallic surfaces intended for implantation.

5.2. History

The first works that we completed on implantable biomaterials date back to the 1980s [MIG 88a, b, c]. They focused on functionalization through radical grafting of the internal surface of tubular materials in order to impart to them coagulant properties which are analogous to those of heparin. This "heparin-like" tubing was functionalized by chemical groups present in heparin molecules – sulfonate (SO_3^-) and carboxylate (COO^-) – in a multi-step process: the grafting of polystyrene onto the internal surface of the tubing was carried out at the CEA (Saclay) by gamma rays of Co 60 and the grafted polystyrene chains were functionalized through chemical substitution. The tubing

Chapter written by Véronique MIGONNEY.

functionalized by SO_3^- and/or COO^- groups was tested *in vitro* and *in vivo* and showed catalytic activity toward the inhibition reaction of thrombin by antithrombin, analogous to that of heparin (i.e. acceleration of the speed of the inhibition reaction of thrombin by antithrombin). The updated mechanism showed that the catalytic activity of the internal surface of the tubular materials was due to a specific adsorption of antithrombin on this "heparin-like" functionalized surface, depending on an antithrombin conformation such that it led to the formation of "thrombin-antithrombin" complexes which could be quantitatively and specifically desorbed by antithrombin from circulating plasmatic antithrombin [MIG 88a, MIG 88b, MIG 88c, CAI 88]. These initial works highlighted the importance of the control of surface chemistry and the choice of functional groups for directing the *in vitro* and *in vivo* biological response.

Works have continued in this area in order to identify, propose and demonstrate the concept of "bioactive" functionalized polymers. With this aim, we synthesized polymers [BEL 00, HEL 02] that have been qualified as "model polymers" since they allow us to understand the origin and the mechanism of the observed biological activities and link them to synthesized macromolecular structures.

Since 2000, the concept of "bioactive polymer" has been extended to polymer or metallic prosthetic surfaces in order to elaborate implantable "bioactive" surfaces. We have established radical grafting techniques which allow for the radical polymerization of "bioactive" homo and co polymers – exhibiting different biological activities directly onto the surface of already commercialized prosthesis surfaces: synthetic ligament (collaboration with the Lars society – Arc sur Tille, France) and total hip prosthesis (collaboration with the Ceraver society – Roissy en France, France).

Several works on "bioactive" polymers and "bioactive" prosthetic surfaces are summarized in the rest of this chapter.

5.3. Model "bioactive" polymers

5.3.1. *Introduction*

Polymers functionalized by SO_3^- and/or COO^- groups have been synthesized by radical polymerization [BEL 00, HEL 02]. These synthesized "model" polymers have helped highlight and understand the biological activities which they induce and also establish a relationship between the biological properties and synthesized macromolecular structures.

The method used for synthesizing the first model bioactive polymers was radical synthesis through homopolymerization and/or copolymerization of chosen monomers:

1) sodium styrene sulfonate (NaSS) and/or methacrylic acid (MA) for anionic functions and the biological activities they can exhibit [MIG 88a, b, c];

2) other monomers for their mechanical or shaping properties such as vinyl chloride (VC), methyl methacrylate (MMA) and hydroxyethyl methacrylate (HEMA).

The aim of these works is to obtain macromolecules and/or materials exhibiting functional groups of interest (SO_3^- and/or COO^-) and exhibiting mechanical properties which would be useful for future applications such as poly vinylchloride (PVC) to prepare heparin-like tubing, polymethyl methacrylate (PMMA) or silicone to elaborate intraocular lenses (IOL), but most importantly which would enable the control of the biological response when they are placed in a biological environment:

– pure proteins: fibronectin (Fn), fibrinogen (Fg), albumin (Alb), or a mix;

– plasma or serum depleted or not in Fn and vitronectin (Vn);

– cells (lines and primary): human and mouse fibroblasts, human and mouse osteoblasts, epithelial corneal cells (collaboration Pr JM Legeais);

– bacteria involved in foreign body infections (commercial strains, model strains and clinically isolated strains – collaboration Pr P Vaudaux, Pr AC Crémieux: *Staphylococcus aureus* (SA), methicillin-resistant *Staphyloccus aureus* (MRSA), *Staphylococcus epidermidis* (SE).

During the entirety of this period, the experiments were carried out *in vitro* and had the aim of highlighting [ELK 02, BER 02, LAT 03, LEG 03, EVA 04, LEG 05] the concept of "bioactive polymer" or "bioactive surface".

Among the biological properties of interest of these "bioactive polymers", we have shown the following:

a) *The modulation of the cell proliferation of fibroblasts and osteoblasts*: during the radical synthesis of copolymers and by varying the proportion of NaSS, MA and MMA monomers, it is possible to obtain functionalized polymer films capable of modulating the proliferation of cells and also controlling cell differentiation [ELK 02, ANA 06].

The mechanisms which form the basis for these activities have been highlighted and come from the selectivity of the adsorption of binding proteins such as Fn and its structure on functionalized surfaces which is modulated according to the chemical composition of the polymer in functional groups SO_3^- and COO^-: it has been shown that it is the differences in exposure of Fn domains to the $\alpha_5\beta_1$ integrins of cells which leads to differences in cell signaling and different proliferations [LAT 03, EVA 04]. It is worthy of note that these properties are visible from the cell adhesion phase – differences in cell adhesion strength, increased focal adhesion points [LAT 03, LEG 05].

It is therefore possible to prevent cell proliferation or improve it by varying the proportions of the functional groups SO_3^- and COO^- present on a surface. These polymers displaying anionic groups are not cytotoxic, but they can be "cytostatic" for certain defined compositions and towards certain cell lines which can be interesting for the development of IOLs [HEL 05, YAM 05] able to prevent secondary cataracts (collaboration with Pr JM Legeais – Hôtel Dieu). On the other hand, it is important to note that for particular chemical compositions they can improve osteoblastic differentiation [ANA 05].

b) *The modulation of bacterial adhesion*: the results obtained for eukaryotic cells and the importance of the selectivity of the protein response in the cell behavior have naturally led to the study of the bacterial response induced by these functionalized surfaces. The main reasons are as follows:

- P. Vaudaux has shown that heparin and dextran when functionalized by SO_4^{2-} and/or COO^- groups displayed inhibiting properties of bacterial adhesion. He also showed the importance and the major role of Fn in the adhesion of *S. aureus* on surfaces such as PMMA and titanium (Ti) [SIN 99, FRA 99, FRA 00, FRA 97];

- PMMA and PVC surfaces functionalized by SO_3^- and/or COO^- groups can be qualified as "heparin-like" or "glycosaminoglycan-like" since they exhibit anionic groups such as SO_3^- and/or COO^- that are present in the heparin molecule and in glycosaminoglycan.

The adsorption of Fn on PMMA surfaces functionalized by the SO_3^- and/or COO^- groups leads to a modulation of the Fn conformation, which is the cause of the modulation of the cell response through possible interactions or non-interactions with the $\alpha_5\beta1$ integrins. In the case of bacteria, the cell receptors involved are "adhesins" (membrane receptors of specific bacteria of binding proteins such as Fn); if the Fn/adhesin interactions are blocked, then bacterial adhesion can be inhibited [BER 02, LAT 03].

The challenge was to show that the SO_3^- and/or COO^- groups present and immobilized at the surface of the "bioactive" polymers (functionalized PMMA and PVC) were capable of causing the same type of bacterial anti-adhesion activity as when they are present on heparin and/or functionalized dextran molecules, polymers that are very different since they are soluble in a physiological environment.

PMMA- or PVC-based copolymers, exhibiting SO_3^- and/or COO^- groups, have been synthesized and tested in the presence of bacteria involved in joint infections, in the presence or absence of Fn, Fg, plasma and/or serum: *S. aureus* model and clinical strains.

This set of works has shown that:

– bacterial adhesion can be modulated by the chemical composition of the surface in SO_3^- and/or COO^- groups [BER 02];

– the proteins present on the surface of these polymers had an important role in bacterial adhesion [EVA 04];

– the main purpose of Fn was confirmed in the adhesion of *S. aureus* to surfaces.

Only the presence of sulfonate groups on the surface of bioactive polymers increases the inhibition properties of the bacterial adhesion of *S. aureus* or MRSA [BER 02].

The results of these biological studies obtained *in vitro* on PMMA- or PVC-based "bioactive" model polymers have led to the continuation of this study by grafting these polymers onto prosthetic polymer or metallic surfaces in order to test them *in vivo* and propose "bioactive" implants or prostheses which are well or bio integrated into biological tissue and/or prevent infections on the prostheses.

5.4. "Bioactive" prosthetic surfaces

5.4.1. *Introduction*

In order to obtain bioactive prosthetic surfaces, "bioactive" polymers were grafted onto the surface of prostheses or implants. The biological activities which were initially observed *in vitro* on model polymers have been confirmed first *in vitro* on surfaces, then *in vivo* on prosthetic surfaces grafted with bioactive polymers.

Two types of works were carried out: (1) research in "macromolecular chemistry" with the development of *bioactive anionic polymer grafting* techniques onto a polymer or metallic surface and (2) research in "biology" and "translational science" with the evaluation of the *in vitro* and *in vivo* biological responses of surfaces grafted with "bioactive" polymers, the validation of the concept and animal models as well as the applications.

The summary of the works undertaken for prosthesis applications is split into two parts:

1) concept and feasibility of the grafting of "bioactive" polymers onto prosthetic surfaces;

2) applications: (a) grafting bioactive polymers onto ligament prostheses from the LARS society; (b) grafting bioactive polymers onto total hip prostheses from the Ceraver society.

5.4.2. *Concept and feasibility of the grafting of "bioactive" polymers onto prosthetic surfaces*

5.4.2.1. *Concept*

The aim was to develop a covalent grafting technique of "bioactive" polymers which would be feasible regardless of the surface of the implant (metallic or polymer), which would be sufficient (grafting rate) in order to obtain the desired biological activity, which would resist the sterilization process of medical tools, and which would be reproducible and applicable on an industrial scale.

In order to complete these objectives, three surfaces were favored:

– silicone in order to show the feasibility of the covalent grafting of a bioactive polymer and the creation of an *in vitro* and *in vivo* biological response;

– poly(ethylene terephthalate) (PET) for the development of the grafting "from" technique and its application in "ligament prostheses";

– Ti or Ti alloy (TA6V) for the development of the covalent grafting of a polymer onto a metallic surface and its application in "total hip prostheses".

Study on feasibility: grafting onto silicone.

The first grafting of "bioactive polymers" onto the surfaces of silicone implants was carried out by Hélary *et al.* on toe prostheses by using a grafting onto technique [YAM 05]. Silicone films were grafted with bioactive polymers in order to validate the technique of silicone surface functionalization and to confirm the properties of the functionalized films [BER 02]; silicone implants were grafted with "bioactive" polymers then implanted into a rabbit within an animal infection model [CRE 03]. The *in vivo* results confirmed the obtained *in vitro* results, first on model polymers and then on functionalized silicone films. The inhibitive properties of the MRSA bacterial adhesion delivered by bioactive polymers when they are grafted onto a surface is preserved regardless of the conditions. The experiments on a rabbit animal model have shown a decrease of the 2 log of the bacteria adhesion/infection in comparison to the control prosthesis (silicone). Furthermore, *ex vivo* experiments – implantations of grafted and non-grafted prostheses followed by its incubation with a bacterial strain – have helped to confirm the importance of the selectivity of the layer of proteins adsorbed on bioactive surfaces towards the bacterial response.

These works on silicone surfaces have allowed us to establish the feasibility of the covalent grafting of a bioactive polymer on the surface of an implant and the creation of the expected biological properties of these surfaces (inhibition of bacterial adhesion).

The drawback to this grafting technique carried out on silicone prostheses was that it is not applicable to implantable surfaces made up of PET, Ti or the alloy TA6V.

In order to overcome this major drawback, we used a radical grafting technique, which allowed for the grafting of "bioactive" polymers directly onto the surface of already commercialized prostheses such as prosthesis synthetic ligament or hip prosthesis. The grafting of the one or more bioactive polymers is carried out using the grafting "from" technique in which the polymerization of the "bioactive" polymer is initiated "from" the surface. This technique requires an activation of the polymer or metallic surface such that it creates active sites – in this case, free radicals from which the radical polymerization process can take place. This grafting procedure was first developed for PET surfaces [BRU 03] and then extended to Ti and TA6V [MIG].

5.4.3. *Applications: (a) grafting of "bioactive" polymers onto LARS ligament prostheses and (b) grafting of "bioactive" polymers onto Ceraver total hip prostheses*

5.4.3.1. *Poly(ethylene terephtalate) (PET) ligament prostheses: grafting of a bioactive polymer and in vitro and in vivo evaluations*

The studies consisted of:

– chemically modifying the surface of the PET ligament prosthesis (LARS society) through the radical grafting of a bioactive polymer. The grafting of poly(sodium styrene sulfonate) (pNaSS) and poly (methacrylic acid) (pMA) onto PET has been developed on films, tissues, fibers and LARS ligament prostheses [BRU 03, CIO 06, PAV 07];

– studying the *in vitro* and *in vivo* biological and biomechanical responses in large animal models (sheep and dogs) [PAV 08, ZHO 07a, PAV 07, ZHO 07b, VIA 11, VIA 13, VAQ 13].

1) Chemistry part

The results on PET samples (films, fibers, ligament) have shown that the radical grafting performed in two steps – ozonation followed by a thermal decomposition of peroxides to create active species on the surface and radical polymerization itself – is perfectly reproducible regardless of the type of sample [CIO 06, PAV 07].

2) Biological part

Studies carried out *in vitro* on film, fabric and fiber samples have highlighted a perfectly controlled cellular response on PET surfaces when they are grafted with bioactive polymers [ZHO 07a and b, VIA 13]:

– primary fibroblastic cells (native ligament) and/or cell lines adhere noticeably better and much more homogeneously on surfaces grafted with bioactive polymers compared to non-grafted surfaces;

– the adhesion of cells is mediated by collagen – the latter is necessary to the interaction between cell and grafted surface, and consequently in the mechanism of this activity.

The inflammatory response is improved on grafted surfaces: the gene expression of TNFα, IL1β, MMP 1 and 13 are under expressed.

Implantations of ligaments grafted with bioactive polymers for periods of 3 months and 1 year in sheep confirm the results obtained *in vitro*. The host response is controlled on grafted ligaments: improved inflammatory reaction and colonization of ligament fibers by fibroblastic cells displaying a similar activity to that of native ligament fibroblasts – molecular biology results 12 months after implantation [VIA 13]. Furthermore, the anchorage of the ligament in the bone section is improved [VAQ 13].

Preclinical assays – implantations carried out on dogs by Professor V. Viateau at the National Veterinary School of Alfort (ENVA France) will confirm the improvement brought by this treatment on the *in vivo* response of ligaments.

To summarize, bioactive ligament prostheses have been the subject of important animal tests:

– on rats: subcutaneous experiments carried out by the Biomatech society;

– on sheep: (>60 animals) [VIA 13, VAQ 13];

– on dogs: carried out within the framework of a preclinical study by V. Viateau.

5.4.3.2. *Titanium alloy TA6V hip prostheses: grafting of a bioactive polymer and in vitro and in vivo evaluations of the biological response*

The aim is to propose a "new generation" of total hip prosthesis made of "bioactive" Ti alloy TA6V, capable of controlling the host response and limiting or even preventing joint infections. This prosthesis is created by grafting the surface with a bioactive polymer coupled with a texturing of the surface. Inflammatory, osteoblastic and bacterial responses are studied *in vitro* and *in vivo* in small animal models.

The development of a bioactive polymer grafting onto Ti or Ti alloy TA6V has been carried out in a multi-step process by using different grafting techniques: (1) "indirect" functionalization through electrochemical oxidation followed by the grafting of a molecule carrying vinyl groups which allows for the initiation of radical polymerization of the bioactive polymer and (2) "direct" functionalization – grafting "from" through chemical or electrochemical oxidation followed by radical polymerization of the bioactive polymer [MIG 06, MAY 08, HEL 08, HEL 09a, MIC 09, HEL 10, KER 10, ALC 13, MIG 13, HEL 09b].

The results of the completed studies have shown that:

– the polymer grafting which is carried out in two steps: oxidation of the surface followed by a thermal decomposition of the peroxides obtained after oxidation in order to create active (radical) species and radical polymerization itself, is reproducible regardless of the type of sample;

– the synthesis conditions in terms of the duration of oxidation and polymerization at 70°C vary depending on the type of oxidation process chosen.

Sterilization does not modify the characteristics of the product.

Studies that have been carried out *in vitro* on samples in the form of plates and disks have highlighted a controlled bacteria and cell response when Ti or Ti alloy is grafted with bioactive polymers [BAY 07, MAY 08, LEC 10, OUG 11, ZOR 11, ALC 13, OUG 13, MIG 13, HEL 09].

The surface grafted with polystyrene sodium sulfonate is biocompatible and does not present any cytotoxicity.

Osteoblast cells (primary Saos, MC3T3, MG63) are more spread out, better adhered and much more homogeneously onto surfaces grafted with bioactive polymers compared to non-grafted polymers. Protein adsorption plays a major role in cell response [FEL 13, FEL 14a, FEL 14b].

Osteoblastic differentiation (ALP and mineralization) is more pronounced on grafted surfaces than on non-grafted ones.

Bacterial adhesion is inhibited when the surfaces are grafted with bioactive polymers: up to 90% compared to a non-grafted surface with an inhibition percentage which depends on the type of proteins present on the surface. Protein adsorption plays a majorly important role in bacterial response, which confirms the results previously obtained with the model polymers by Berlot *et al.* [BER 02].

In vivo experiments on rabbits have allowed for the demonstration of short-term osseointegration of implants grafted with polyNaSS.

The first *in vivo* infection tests carried out on rabbits demonstrate the purpose of grafting on Ti alloy in order to reduce bacterial adhesion/infection: the difference between Ti surfaces grafted with bioactive polymers and non-grafted ones is on the same level as that observed with silicone. These tests were carried out on multiple sets of animals.

5.5. Bibliography

[ALC 13] AL CHEICK H., PAVON G., HÉLARY G., *et al.*, "PolyNaSS grafting on titanium surfaces enhances osteoblast differentiation and inhibits *Staphylococcus aureus* adhesion", *Journal of Materials Science – Materials in Medicine*, vol. 24, no. 7, pp. 1745–54, July 2013.

[ANA 06] ANAGNOSTOU F., DEBET A., PAVON-DJAVID G., *et al.*, "Osteoblast functions on functionalized PMMA-based polymers exhibiting *S. aureus* adhesion inhibition", *Biomaterials*, vol. 27, no. 21, pp. 3912–3919, 2006.

[BAU 07] BAUER T., PAVON-DJAVID G., HÉLARY G., *et al.*, "Inhibition de l'adhérence bactérienne sur le ciment", *Revue de Chirurgie Orthopédique et Réparatrice de l'Appareil Moteur*, vol. 93, no. 7-Suppl.1, pp. 156–156, 2007.

[BEL 00] BELLENEY J., HÉLARY G., MIGONNEY V., "Terpolymerization of 3-methacryloxypropyl tris(trimethylsiloxy)silane, methacrylic acid and dimethyl octyl ammonium styrene sulfonate: determination of the reactivity ratios", *European Polymer Journal*, vol. 36, no. 11, pp. 2365–2369, 2000.

[BER 02] BERLOT S., PAVON G., HELARY G., *et al.*, "Biomimetic PMMA based terpolymers: modulation of bacterial adhesion effect chemical composition and monomers distribution effect", *Biomacromolecules*, vol. 3, no. 1, pp. 63–68, 2002.

[BRU 03] BRULEZ B., LABOUREAU J.P., MIGONNEY V., *et al.*, Procédé de traitement de surfaces polymères, Biomimetic prosthetic ligament and production method thereof, European Patent No. 1,599,238 (2006); US Patent No. 7,700,147 (2010); WO Patent No. 2,004,067,051 (2004), French Patent No. 0300495, 2003.

[CAI 88] CAIX J., MIGONNEY V., BAQUEY A., *et al.*, "Control and isotopic quantification of affinity of antithrombin III for heparin-like surfaces", *Biomaterials*, vol. 9, pp. 62–65, 1988.

[CIO 06] CIOBANU M., GUEGUEN V., SIOV A., *et al.*, "Radical graft polymerization of styrene sulfonate grafting on poly (ethylene terephthalate) films for ACL applications: 'grafting from' and chemical characterization", *Biomacromolecules*, vol. 7, no. 3, pp. 755–760, 2006.

[CRÉ 03] CRÉMIEUX A.-C., PAVON-DJAVID G., SALEH MGHIR A., *et al.*, "Bioactive polymers grafted on silicone to prevent *Staphylococcus aureus* prosthesis adherence: in vitro and in vivo study", *Journal of Applied Biomaterials & Biomechanics*, vol. 1, pp. 178–185, 2003.

[ELK 02] EL KADHALI F., HELARY G., MIGONNEY V., *et al.*, "Modulation of fibroblast cell proliferation with functionalized PMMA based copolymers: chemical composition and monomers distribution effect", *Biomacromolecules*, vol. 3, no. 1, pp. 51–56, 2002.

[EVA 04] EVANS M.D.M., PAVON-DJAVID G., HÉLARY G., *et al.*, "Vitronectin is significant in the adhesion of lens epithelial cells to PMMA polymers", *J Biomed Mater Res*, vol. 69A, pp. 469–476, 2004.

[FEL 13] FELGUEIRAS H., MIGONNEY V., "Presence of sulfonate groups on Ti6Al4V surfaces enhances osteoblastic attachment strength at the interf", *IRBM*, vol. 34, no. 6, pp. 371–375, December 2013.

[FEL 14a] FELGUEIRAS H., MIGONNEY V., "Sulfonate groups grafted on Ti6Al4V favor MC3T3-E1 cell performance in serum free medium conditions", *Mater Sci Eng C Mater Biol Appl.*, no. 39, pp. 196–202, 1 June 2014.

[FEL 14b] FELGUEIRAS H.P., SOMMERFELD S.D., MURTHY N.S., *et al.*, "Poly(NaSS) Functionalization Modulates the Conformation of Fibronectin and Collagen Type I To Enhance Osteoblastic Cell Attachment onto Ti6AlV", *Langmuir*, vol. 30, no. 31, pp. 9477–83, 12 August 2014.

[FRA 99] FRANÇOIS P., LETOURNEUR D., LEW D.P., *et al.*, "Inhibition by heparin and derivatized dextrans of Staphylococcus epidermidis adhesion to in vitro fibronectin-coated or explanted polymer surfaces", *J Biomater Sci Polym Ed.*, vol. 10, no. 12, pp. 1207–1221, 1999.

[FRA 00] FRANÇOIS P., SCHRENZEL J., STOERMAN-CHOPARD C., *et al.*, "Identification of plasma proteins adsorbed on hemodialysis tubing that promote Staphylococcus aureus adhesion", *J Lab Clin Med.*, vol. 135, no. 1, pp. 32–42, January 2000.

[FRA 97] FRANÇOIS P., VAUDAUX P., TABORELLI M., *et al.*, "Influence of surface treatments developed for oral implants on the physical and biological properties of titanium. (II) Adsorption isotherms and biological activity of immobilized fibronectin", *Clin Oral Implants Res.*, vol. 8, no. 3, pp. 217–225, June 1997.

[HEL 02] HÉLARY G., BELLENEY J., MIGONNEY V., "Terpolymerization of methyl methacrylate, poly(ethylene glycol) methyl ether methacrylate or poly(ethylene glycol) ethyl ether methacrylate with methacrylic acid and sodium styrene sulfonate: determination of the reactivity ratios", *European Polymer Journal*, vol. 38, no. 3, pp. 439–444, 2002.

[HEL 09a] HÉLARY G., MIGONNEY V., "Bioactive polymers as coating to elaborate biomaterials exhibiting bone integration properties", in PLANELL J.A., BEST S., LACROIX D., *et al.*, (eds.), *Bone Repair Biomaterials*, Woodhead Publishing, pp. 309–323, 2009.

[HEL 09b] HÉLARY G., NOIRCLÈRE F., MAYINGI J., *et al.*, "A new approach to graft bioactive polymer on titanium implants: improvement of MG 63 cell differentiation onto this coating", *Acta Biomaterialia*, vol. 5, no. 1, pp. 124–133, 2009.

[HEL 10] HÉLARY G., NOIRCLÈRE F., MAYINGI J., *et al.*, "A bioactive polymer grafted on titanium oxide layer obtained by electrochemical oxidation – improvement of cell response", *Journal of Materials Science – Materials in Medicine*, vol. 21, no. 2, pp. 655–663, 2010.

[HEL 08] HÉLARY G., POUSSARD L., ZREIQAT H., *et al.*, "Functionalization of biomaterials for joint implant application", *Bio-Medical Materials and Engineering*, vol. 18, nos. 4–5, pp. 237–239, 2008.

[HEL 05] HELARY G., YAMMINE P., MIGONNEY V., "Surface modification of hydrogel intraocular lenses to prevent cell proliferation", *Journal of Applied Biomaterials & Biomechanics*, vol. 2, pp. 183–190, 2005.

[KER 10] KERNER S., MIGONNEY V., PAVON G., *et al.*, "Bone tissue response to titanium implant surfaces modified with carboxylate and sulfonate groups", *J Mater Sci: Mater Med*, vol. 21, pp. 707–715, 2010.

[LAT 03] LATZ C., PAVON-DJAVID G., HÉLARY G., *et al.*, "Alternative intracellular signaling mechanism involved in the inhibitory biological response of functionalized PMMA-based terpolymers", *Biomacromolecules*, vol. 4, no. 3, pp. 766–771, 2003.

[LEC 10] LECOMTE M., WENGER F., PONTHIAUX P., *et al.*, "Modélisation de l'effet de la rugosité sur l'adhésion d'ostéoblastes: application au titane", *Matériaux & Techniques*, vol. 98, pp. 49–57, 2010.

[LEG 03] LE GUILLOU-BUFFELLO D., HÉLARY G., GINDRE M., *et al.*, "Measurement of the inhibition of cell processes on five polymeric films by quartz crystal resonator; comparison with the cell counting technique", *Proceedings WCU*, pp. 267–270, 2003.

[LEG 04] LE GUILLOU-BUFFELLO D., HÉLARY G., GINDRE M., *et al.*, "Estimation de l'inhibition de l'adhésion cellulaire sur des films de polymère par la technique du résonateur à quartz", *ITBM-RBM*, vol. 2, pp. 2534–2539, 2004.

[LEG 05] LE GUILLOU-BUFFELLO D., HÉLARY G., GINDRE M., *et al.*, "Monitoring cell adhesion processes on bioactive polymers with the quartz crystal resonator technique", *Biomaterials*, vol. 26, no. 19, pp. 4197–4205, 2005.

[MAY 08] MAYINGI J., HÉLARY G., NOIRCLÈRE F., *et al.*, "Grafting of bioactive polymers onto titanium surfaces and human osteoblasts response", *IRBM*, vol. 29, no. 1, pp. 1–6, 2008.

[MIC 10] MICHIARDI A., HÉLARY G., NGUYEN P.-C.T., *et al.*, "Bioactive polymer grafting onto titanium alloy surfaces", *Acta Biomater.*, vol. 6, no. 2, pp. 667–675, 2010.

[MIG 88a] MIGONNEY V., FOUGNOT C., JOZEFOWICZ M., "Heparin-like tubings. I. Preparation, characterization and biological in vitro activity assessment", *Biomaterials*, vol. 9, pp. 145–149, 1988.

[MIG 88b] MIGONNEY V., FOUGNOT C., JOZEFOWICZ M., "Heparin-like tubings. II. Mechanism of the thrombin-antithrombin III reaction at the surface", *Biomaterials*, vol. 9, pp. 230–234, 1988.

[MIG 88c] MIGONNEY V., FOUGNOT C., JOZEFOWICZ M., "Heparin-like tubings: III. Kinetics and mechanism of thrombin, antithrombin III and thrombin-antithrombin complex adsorption under controlled-flow conditions", *Biomaterials*, vol. 9, no. 5, pp. 413–418, 1988.

[MIG 13] MIGONNEY V., HÉLARY G., BEN AISSA I., *et al.*, "ACTISURF", *IRBM*, vol. 34, no. 2, pp. 180–185, 2013.

[MIG 06] MIGONNEY V., NOIRCLÈRE F., HÉLARY G., Procédé de greffage de polymères bioactifs sur des matériaux prothétiques, French Patent No. FR 06 05 073, June 2006.

[OUG 11] OUGHLIS S., LESSIM S., CHANGOTADE S., *et al.*, "Development of proteomic tools to study protein adsorption on a biomaterial, titanium grafted with poly (sodium styrene sulfonate)", *Chromatogr B Analyt Technol Biomed Life Sci.*, vol. 879, no. 31, pp. 3681–3687, 2011.

[OUG 13] OUGHLIS S., LESSIM S., CHANGOTADE S., *et al.*, *J. Biomed Mat Res*, 2013.

[PAV 07] PAVON-DJAVID G., GAMBLE L.J., CIOBANU M., *et al.*, "Bioactive poly(ethylene terephtalate) fibers and fabrics: grafting, chemical characterization, and biological assessment", *Biomacromolecules*, vol. 8, no. 11, pp. 3317–3325, 2007.

[PAV 08] PAVON-DJAVID G., ISAC N., VIATEAU V., *et al.*, "Observation de ligaments artificiels de genou après implantation en MEB à pression variable", *IRBM*, vol. 29, no. 6, pp. 380–384, 2008.

[SIN 99] SINHA B., FRANÇOIS P.P., NÜSSE O., *et al.*, "Fibronectin-binding protein acts as Staphylococcus aureus invasin via fibronectin bridging to integrin alpha5beta1", *Cell Microbiol.*, vol. 1, no. 2, pp. 101–117, September 1999.

[VAQ 13] VAQUETTE C., VIATEAU V., GUÉRARD S., *et al.*, "The effect of polystyrene sodium sulfonate grafting on polyethylene terephthalate artificial ligaments on in vitro mineralisation and in vivo bone tissue integration", *Biomaterials*, vol. 34, no. 29, pp. 7048–7063, September 2013.

[VIA 13] VIATEAU V., MANASSERO M., ANAGNOSTOU F., *et al.*, "Biological and biomechanical evaluation of the ligament advanced reinforcement system (LARS AC) in a sheep model of anterior cruciate ligament replacement: a 3-month and 12-month study", *Arthroscopy*, vol. 29, no. 6, pp. 1079–1088, 2013.

[VIA 11] VIATEAU V., ZHOU J., GUÉRARD S., *et al.*, "Ligart: ligament synthétique 'bioactif' et 'biointégrable' permettant la réhabilitation rapide du patient: greffage chimique, évaluations biologiques in vitro et vivo, experimentation animale, etude préclinique", *IRBM*, vol. 32, no. 2, pp. 118–122, 2011.

[YAM 05] YAMMINE P., PAVON-DJAVID G., HÉLARY G., *et al.*, "Surface modification of silicone intraocular implants to inhibit cell proliferation", *Biomacromolecules*, vol. 6, no. 5, pp. 2630–2637, 2005.

[ZHO 07a] ZHOU J., CIOBANU M., PAVON-DJAVID G., *et al.*, "Morphology and adhesion of human fibroblast cells cultured on bioactive polymer grafted ligament prosthesis", *Conf Proc IEEE Eng Med Biol Soc.*, vol. 1–16, pp. 5115–5118, 2007.

[ZHO 07b] ZHOU J., PAVON-DJAVID G., ANAGNOSTOU F., *et al.*, "Inhibition de l'adhérence de Porphyromonas gingivalis sur la surface de titane greffé de poly(styrène sulfonate de sodium)", *IRBM*, vol. 28, no. 1, pp. 42–48, 2007.

[ZOR 11] ZORN G., BAIO J.E., WEIDNER T., *et al.*, "Characterization of poly(sodium styrene sulfonate) thin films grafted from functionalized titanium surfaces", *Langmuir*, vol. 27, no. 21, pp. 13104–13112, 2011.

6

Functionalization of Biomaterials and Applications

6.1. Introduction

Bacterial adhesion, biocompatibility and biointegration on implanted prosthetic materials represent major problems for public health. To combat these problems, solid surface modification is required. An alternative that is becoming more and more recurrent nowadays is the grafting of biomolecules or/and bioactive polymers. The use bioactive polymers has been shown to be an excellent solution. Tethering bioactive polymers to a surface has emerged as a promising tool to tailor surface properties. Polymer brushes can be immobilized on appropriate surfaces using either a physisorption (chain attachment mainly through van der Waals interactions) or a covalent strategy (anchoring by chemical bonds). Although the physisorption technique has been used to modify polymers and metallic surfaces, they suffer from the drawbacks inherent in their non-permanent nature: the release of polymers from such modified surfaces and the subsequent loss of activity potentially make them unsuitable for most biomedical applications. So, it is important for polymers to be covalently immobilized on the surface. Alternatively, several techniques for covalently tethering well-defined polymer brushes onto surfaces have been developed, including the covalent attachment of end-functionalized polymers incorporating an

Chapter written by Céline FALENTIN-DAUDRE.

appropriate anchor ("grafting to" Figure 6.1) or the *in situ* polymerization initiated from the surface ("grafting from" Figure 6.2) [UYA 98, MIN 08].

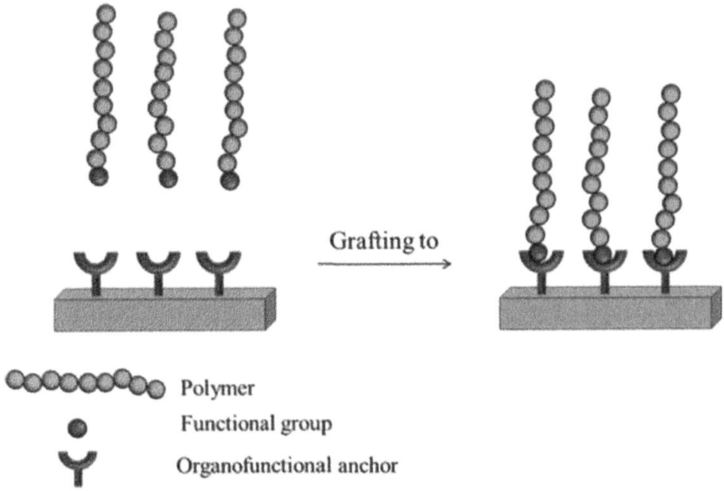

Figure 6.1. *Schematic description of the "grafting to" process*

The "grafting to" method (Figure 6.1) involves reaction of (end)-functionalized polymer molecules with complementary functional groups located on the surface, resulting in the formation of tethered chains. If a surface that is to be modified does not possess the necessary functionalities, several techniques can be used for initial surface functionalization or to immobilize anchors onto the solid surface, including plasma treatments, self-assembled monolayer deposition or chemisorption of a thin layer of reactive polymers. End-functionalized polymers with a narrow molecular weight distribution can be synthesized by living anionic, cationic, radical, group transfer and ring opening metathesis polymerizations. Afterward, the functional groups of polymers can be involved in further chemical reactions to attach polymers on anchors [UYA 98, MIN 08, ZDY 11].

The "grafting from" (Figure 6.2) approach has attracted considerable attention in recent years in the preparation of tethered polymers on a solid substrate surface. The initiators are immobilized onto the surface followed by *in situ* surface-initiated polymerization (non-controlled or controlled polymerization) generating tethered polymers. The immobilization of initiators on the substrate surface can be achieved with plasma, UV irradiation or ozone treatment of the surface [UYA 98, MIN 08].

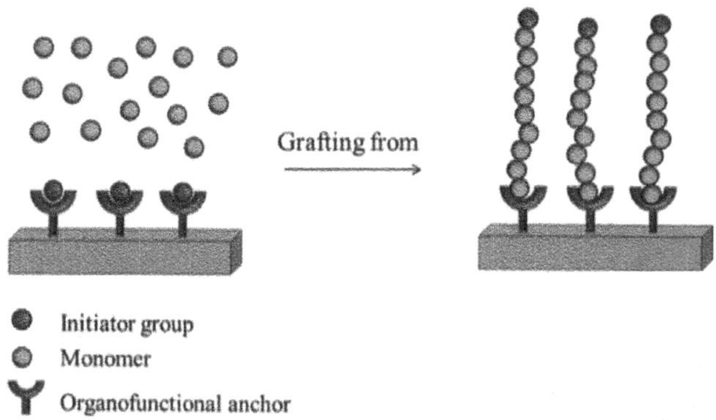

Figure 6.2. *Schematic description of the "grafting from" process*

6.2. Applications

6.2.1. *"Grafting to" on stainless steel surfaces for antibacterial and antiadhesion properties*

Stainless steel (SS) is widely used in daily life because of its resistance to corrosion and chemicals, and its mechanical and esthetic properties. It can be found in many areas such as medical and household appliances, and the building and food industries [HEL 98]. Despite good cleanability, bacteria absorb easily on stainless steel. On SS, bacteria form colonies and subsequently biofilms that serve as reservoirs for the development of pathogenic infection [DON 02]. Therefore, for hygienic reasons, it is necessary to develop new strategies to protect the surface against these microorganisms by preventing them from adhering and, in the case of adhesion, killing them while removing the biolfilm.

This can be accomplished by physical or chemical surface modifications. To avoid the colonization of the surface by bacteria, two principal strategies have been developed: the first one is to develop antibacterial surface films [IGN 09, LEE 04, CHA 09, FAL 12, GLI 09, FAU 11] and the second one is to immobilize antifouling or antiadhesion coatings that prevent bacterial adhesion [OST 01, STA 08, BOU 04, DON 07, ZHA 01, CAR 09]. Antibacterial surfaces can be generated by painting the substrates with biocide-loaded coatings [HET 06] or by chemically grafting the biocides to the surfaces. In the first case, the antibacterial activity is due to the diffusion of the biocide out of the coating. In the second case, bacteria are killed when in contact with the surface. Other examples of antibacterial coating onto SS can be found in the literature on electrografting of acrylates post-modified to obtain antibacterial properties [IGN 09], the grafting of macromolecules bearing quaternary ammonium groups [LEE 04], the formation of biocidal multilayered polyelectrolyte films [CHA 09, FAL 12] and the grafting of peptides [GLI 09, FAU 11]. For antifouling and antiadhesion coatings, many examples can be found such as self-assembled monolayers [OST 01, STA 08], the formation of a multilayer film with poly(ethylene glycol) (PEG) [BOU 04], the grafting of PEG by a cold plasma technique [DON 07] or by silane coupling agents [ZHA 01] and the grafting of lysozymes [CAR 09]. However, these are generally multi-step processes and mostly use organic solvents that are toxic and are not desirable for industrial applications.

In order to impart long-term durability to the coating and prevent biofilm formation when SS is aging, Falentin *et al.* [FAU 12] report the covalent anchoring of (bio)molecules onto SS surfaces via the "grafting to" method (Figure 6.3). The grafting to method for functionalizing SS surfaces could be based on the use of a readily accessible anchor incorporating both an anchoring group that is capable of forming, under mild condition, a robust stable layer and a reactive function facilitating the modular and efficient post-functionalization of SS surfaces. Here, two (bio)molecules were studied: Dispersin and PEG. Dispersin B (Dsp B) is an enzyme active against *N*-acetylglucosamine containing extracellular polysaccharides that are a part of biofilms. It consequently degrades the polysaccharides that the bioorganisms use to anchor and colonize the substrates [RAM 05]. Furthermore, this water-soluble enzyme is highly active against both Gram negative [ITO 05] and positive [KAP 04] bacteria such as *E. coli* and *S. epidermidis*, respectively. PEG is one of the most commonly used synthetic polymers to impart protein resistance to a surface [BAN 11].

In this work, long-term antibacterial, antiadhesion and antibiofilm activities are afforded to industrial SS surfaces following a green and bio-inspired strategy (Figure 6.3). Starting from catechol-bearing synthetic polymers, the film cross-linking and the grafting of active (bio)molecules are possible under environmentally friendly conditions (in aqueous media and at room temperature). A bio-inspired polyelectrolyte, a polycation-bearing catechol, is used as the film-anchoring polymer while poly(methacrylamide)-bearing quinone groups serve as the cross-linking agent in combination with a polymer-bearing primary amine groups. The amine/quinone reaction is exploited to prepare a stable solution of nanogels in water at room temperature that can be easily deposited to SS. This coating provides quinone-functionalized surfaces that are then used to covalently anchor active (bio)molecules (antibiofilm enzyme and antiadhesion polymer) through thiol/quinone reactions [FAU 12]. For that purpose, thiol end-functionalized Dsp B and PEG-SH that contains two external cysteines (and thus thiol groups) in its sequence were considered for the conjugation.

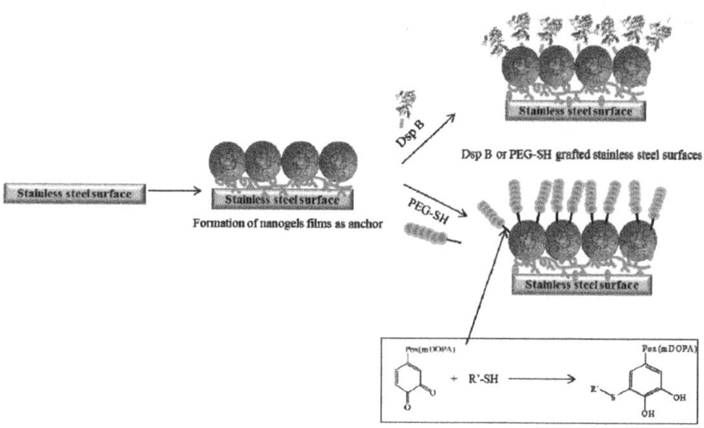

Figure 6.3. *Schematic description of the "grafting to" process onto stainless steel surfaces*

Impact of grafted Dsp B on *S. epidermidis* biofilm was then quantified. The number of viable *S. epidermidis* adherent bacteria was decreased by 86% (standard deviation: 9%) on Dsp B-coated surfaces compared to their uncoated counterparts [FAU 12]. Immobilized PEG-SH coatings were also active against *S. epidermidis* as traduced by a 93% (standard deviation: 4%) decrease of the adherent bacteria compared with the uncoated counterparts [FAU 12]. This clearly showed the benefit of the permanent immobilization of the active molecules on the surfaces.

6.2.2. *Grafting of bioactive polymers onto titanium implants*

Titanium and its alloys are widely used in orthopedic and dental implants for its excellent resistance to corrosion and its biocompatibility. In spite of this property, insufficient integration into surrounding bone often occurs. An uncontrolled inflammation process inducing fibrous capsule formation prevents the generation of a stable implant to host tissue binding and consequently implants can fail under shear stress, requiring revision surgery [KLA 01]. Over the last 15 years, many studies have been done on increasing the osteointegration by modifying surface properties (roughness, topography, surface charges, passivation and wettability) using different methods, such as mechanical [BAN 06, BIG 02], chemical [WAN 04], thermal [PAR 07, SAL 07] or electrochemical treatments [SUL 02, LEG 07]. However, even if modified surfaces have a higher early level of cell attachment than untreated titanium surface, the successful implantation rate is not satisfying. Biochemical methods of surface modification are promising approaches [REY 07, PAL 05, POR 04]. The aim is to control the tissue–implant interface by the immobilization of proteins, enzymes or peptides for the purpose of inducing specific cell responses. The main difficulty is to ensure the stability of the biomolecules binding to the surface of the implant and its accessibility to active sites of cells. Physical adsorption is not successful for long-term implantation mainly due to the desorption of biomolecules. The covalent attachment of biomolecules to the titanium surface creates the problem of disruption under physiological medium or mechanical stress, which needs to be resolved. Covalent attachment requires the use of different chemical reactions which can be aggressive toward the biomolecules, reducing their potential bioactivity [MUL 06]. Other problems are the cost of the biomolecules, which can be a limitation to industrial application.

In order to improve the long-term osteointegration of titanium surfaces, bioactive polymers bearing ionics groups such as sulfonates (sodium polystyrene sulfonate poly (NaSS)) are grafted in a covalent manner onto titanium surface. Migonney *et al.* [ELK 02, ANA 06] have shown that polymers bearing appropriate chemical functions can modulate the cell attachment and spread onto these bioactive polymers and the cell activity. The distribution of these ionic groups along the macromolecular chains creates active sites which can interact with extracellular proteins, such as fibronectin, involved in cell response. Migonney *et al.* [HEL 10, HEL 09, MIC 10, KER 10, BEN 11, FEL 13, FEL 14] report the grafting of an ionic polymer model poly(sodium styrene sulfonate) (poly (NaSS)) in a two-step reaction procedure (Figure 6.4) onto titanium and alloy titanium surfaces. Treatment of the titanium surface by a mixture of sulfuric acid and hydrogen peroxide helps in the formation of titanium hydroxide and titanium peroxide. In the second reaction step, heating of a metal implant, placed in a concentrated sodium styrene sulfonate monomer (NaSS), induces the decomposition of titanium peroxides with the formation of radicals capable of initiating the polymerization of NaSS.

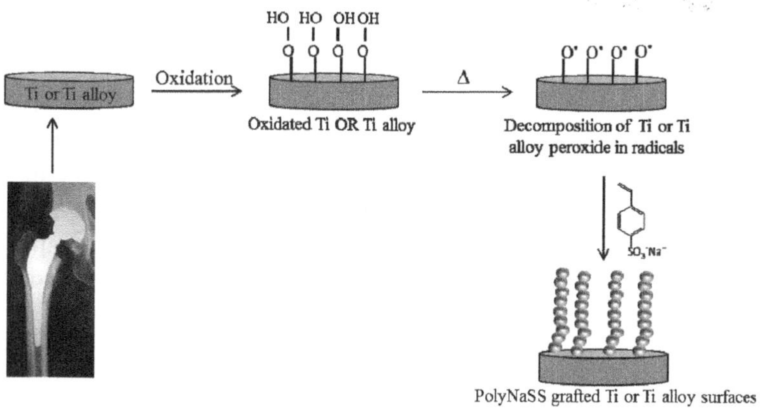

Figure 6.4. *Schematic description of the grafting process onto titanium or titanium alloy implants*

Different techniques (colorimetry, Fourier-transformed infrared spectra recorded in an attenuated total reflexion, X-ray photoelectron spectroscopy technique and contact angle measurements) were used to check the presence of the bioactive polymers at the surface of the titanium samples. The grafting process was found to be successful with the colorimetric method uptake value of 5 $\mu g/cm^2$ for titanium surfaces and for titanium alloy surfaces [HEL 10, FEL 13].

Bacterial studies have shown the advantage of poly(NaSS) grafted onto titanium and onto alloy titanium (Ti6Al4V) surfaces. Bacterial adhesion study showed that titanium graft surfaces exhibited high inhibition of *S. aureus* adhesion at levels greater than 90% and of *S. epidermidis* adhesion at levels greater than 70% when compared to titanium [BEN 11]. Then, osteoblastic cell response was studied on polished, oxidized and grafted titanium and alloy samples. Cell adhesion, alkaline phosphatase activity and calcium nodule formation were significantly enhanced on grafted titanium samples compared to unmodified surfaces [HEL 10, FEL 13, FEL 14]. The bioactive polymer together with the titanium material offers a promising solution for the fast biomaterial osteointegration to be used in the orthopedic and dental field.

6.2.3. *Radical graft polymerization of bioactive polymers on poly(ethylene terephthalate) (PET) for anterior cruciate ligament applications*

Knee ligaments are commonly injured, especially by people participating in sports such as skiing or football. Injury of the anterior cruciate ligament (ACL), in particular, can sometimes result in the termination of an athlete's participation in competition. ACL rupture is the most common sport injury and due to its poor healing capacity [DUT 06, ZAN 06], surgical treatments are often required for restoring the function of the knee.

After an ACL tear, the best solutions for repair are either ligament replacement by a tendon autograft or reconstruction using an artificial ligament. Autologous tissues such as Bone Patellar Tendon Bone [ALM 74] and Semitendinous and Gracilis Tendon are the most used grafts for ACL reconstruction providing excellent clinical results [SAJ 11]. These above-mentioned techniques take advantage of the use of autologous natural tissue which facilitates the tissue integration of implants, but nevertheless they present major inconveniences related to tissue harvesting and therefore induced site morbidity, pain, moderate to long recovery periods before resuming physical exercise and professional activity.

Synthetic ligament failures are often caused by the lack of integration of the ligament in the bone tunnels along with non-optimal cell infiltration and organization within the ligament compartment [LI 12]. Therefore another step forward is the improvement of the artificial ligament bio-integration which is believed to play a significant role in enhancing the long-term performances of these synthetic implants [LI 11]. Therefore, the surface properties of the synthetic ACL must be modified to achieve better bio-integration with the host. Grafting synthetic polymer, biopolymer or peptides onto implant surfaces is shown to significantly affect the cell behavior both *in vitro* and *in vivo*. To this end, Ciobanu *et al.* have developed a radical graft polymerization of polystyrene sodium sulfonate (PolyNaSS) onto PET [CIO 06, PAV 07] in order to enhance the interaction of the artificial ligament with relevant cells. The grafting procedure [CIO 06, PAV 07] is chemically illustrated in Figure 6.5. Radical graft polymerization of the sodium salt of styrene sulfonate (NaSS) onto PET was performed using the "grafting from" approach (Figure 6.5). Prior to the grafting, ligaments were activated by ozonation to generate peroxide and hydroperoxide reactive species on the surface. The radical polymerization of NaSS was initiated by thermal decomposition of the hydroperoxides.

This study [VAQ 13] has shown the beneficial effect of polyNaSS grafting onto the artificial ligament both *in vitro* and *in vivo*. The grafting enhanced ALP secretion in osteoblasts and had a significant impact on their *in vitro* mineralization as mineralized nodules were detected at the surface of the artificial ligament 6 weeks after seeding under osteogenic induction. This effect was translated *in vivo* into an enhanced direct ligament to bone contact and decreased fibrous scar tissue at the bone–ligament interface 12 months after implantation in an ovine ACL reconstruction model. These results provide evidence that PolyNaSS grafting of PET ligament improves artificial ligament osseo-integration within the bone tunnels [VAQ 13].

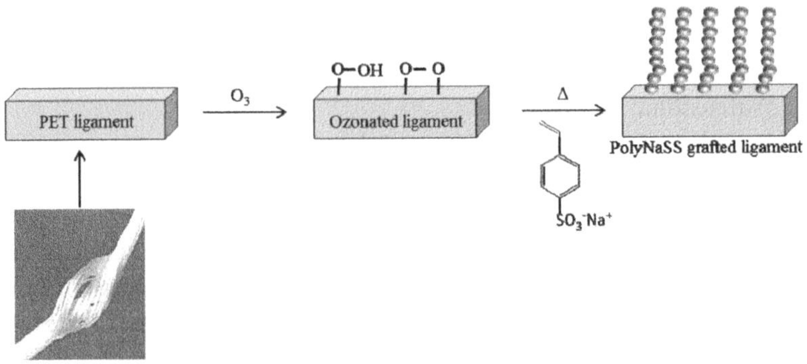

Figure 6.5. *Schematic description of the grafting process on PET ligament*

It is possible to use this method of grafting onto biodegradable polyesters (for example) for use in tissue engineering. This is a very attractive method but it will change the grafting conditions such as the temperature of grafting and solvent conditions. Indeed, the biodegradable polyesters have different properties compared to PET. Although these structures are envisioned as the future prostheses, there is still a lot more research that needs to be conducted before they can be transposed from bench to bedside.

6.3. Bibliography

[BOU 04] BOULMEDAIS F., FRISCH B., ETIENNE O., *et al.*, "Polyelectrolyte multilayer films with pegylated polypeptides as a new type of antimicrobial protection for biomaterials", *Biomaterials*, vol. 25, pp. 2003–2011, 2004.

[CAR 09] CARO A., HUMBLOT V., METHIVIER C., *et al.*, "Grafting of lysozyme and/or poly(ethylene glycol) to prevent biofilm growth on stainless steel surfaces", *J. Phys. Chem. B*, vol. 113, pp. 2101–2109, 2009.

[CHA 09] CHARLOT A., SCIANNAMEA V., LENOIR S., *et al.*, "All-in-one strategy for the fabrication of antimicrobial biomimetic films on stainless steel", *J. Mater. Chem.*, vol. 19, pp. 4117–4125, 2009.

[CIO 06] CIOBANU M., SIOVE A., GUEGUEN V., *et al.*, "Radical graft polymerization of styrene sulfonate on poly(ethylene terephthalate) films for ACL applications: 'Grafting from' and chemical characterization", *Biomacromolecules*, vol. 7, pp. 755–760, 2006.

[DON 07] DONG B., JIANG H., MANOLACHE S., *et al.*, "Plasma-mediated grafting of poly(ethylene glycol) on polyamide and polyester surfaces and evaluation of antifouling ability of modified substrates", *Langmuir*, vol. 23, pp. 7306–7313, 2007.

[DON 02] DONLAN R.M., COSTERTON J.W., "Biofilms: survival mechanisms of clinically relevant microorganisms", *Clin. Microbiol. Rev.*, vol. 15, pp. 167–193, 2002.

[DUT 06] DUTHON V.B., BAREA C., ABRASSART S., *et al.*, "Anatomy of the anterior cruciate ligament", *Knee Surg Sport Traumatol Arthrose*, vol. 14, pp. 203–213, 2006.

[ELK 02] EL KHADALI F., HELARY G., PAVON-DJAVID G., *et al.*, "Modulating fibroblast cell proliferation with functionalized poly(methyl methacrylate) based copolymers: chemical composition and monomer distribution effect", *Biomacromolecules*, vol. 3, pp. 51–56, 2002.

[FAL 12] FALENTIN-DAUDRE C., FAURE E., SVALDO-LANERO T., *et al.*, "Antibacterial polyelectrolyte micelles for coating stainless steel", *Langmuir*, vol. 28, pp. 7233–7241, 2012.

[FAU 11] FAURE E., LECOMTE P., LENOIR S., *et al.*, "Sustainable and bio-inspired chemistry for robust antibacterial activity of stainless steel", *J. Mater. Chem.*, vol. 21, pp. 7901–7904, 2011.

[FAU 12] FAURE E., FALENTIN-DAUDRE C., SVALDO-LANERO T., *et al.*, "Functional nanogels as platforms for imparting antibacterial, antibiofi lm, and antiadhesion activities to stainless steel", *Adv. Funct. Mater.*, vol. 22, no. 24, pp. 5271–5282, 2012.

[FEL 13] FELGUEIRAS H., MIGONNEY V., "Presence of sulfonate groups on Ti6Al4V surfaces enhances osteoblastic attachment strength at the interface", *IRBM*, vol. 34, pp. 371–375, 2013.

[FEL 14] FELGUEIRAS H., MIGONNEY V., "Sulfonate groups grafted on Ti6Al4V favor MC3T3-E1 cell performance in serum free medium conditions", *Materials Science and Engineering: C*, vol. 39, pp. 196–202, 2014.

[GLI 09] GLINEL K., JONAS A.M., JOUENNE T., *et al.*, "Antibacterial and antifouling polymer brushes incorporating antimicrobial peptide", *Bioconjugate Chem.*, vol. 20, pp. 71–77, 2009.

[HEL 09] HÉLARY G., NOIRCLÈRE F., MAYING J., *et al.*, "A new approach to graft bioactive polymer on titanium implants: improvement of MG 63 cell differentiation onto this coating", *Acta Biomater.*, vol. 5, pp. 124–133, 2009.

[HEL 10] HÉLARY G., NOIRCLÈRE F., MAYING J., *et al.*, "A bioactive polymer grafted on titanium oxide layer obtained by electrochemical oxidation. improvement of cell response", *J. Mater. Sci. Mater. Med.*, vol. 21, pp. 655–663, 2010.

[HEL 98] HELSEN J.A., BREME H.J., *Metals as Biomaterials*, Wiley, New York, 1998.

[HET 06] HETRICK E.M., SCHOENFISCH M.H., "Reducing implant-related infections: active release strategies", *Chem. Soc. Rev.*, vol. 35, pp. 780–789, 2006.

[IGN 09] IGNATOVA M., VOCCIA S., GABRIEL S., *et al.*, "Stainless steel grafting of hyperbranched polymer brushes with an antibacterial activity: synthesis, characterization, and properties", *Langmuir*, vol. 25, pp. 891–902, 2009.

[ITO 05] ITOH Y., WANG X., HINNEBUSCH B.J., *et al.*, "Depolymerization of β-1,6-*N*-acetyl-d-glucosamine disrupts the integrity of diverse bacterial biofilms", *J. Bacteriol.*, vol. 187, pp. 382–387, 2005.

[KAP 04] KAPLAN J.B., RAGUNATH C., VELLIYAGOUNDER K., *et al.*, "Enzymatic detachment of *Staphylococcus epidermidis* biofilms", *Antimicrob. Agents Chemother.*, vol. 48, pp. 2633–2336, 2004.

[KER 10] KERNER S., MIGONNEY V., PAVON-DJAVID G., *et al.*, "Bone tissue response to titanium implant surfaces modified with carboxylate and sulfonate groups", *J. Mater. Sci. Mater. Med.*, vol. 21, pp. 707–715, 2010.

[KLA 01] KLADUNE WINDLER M., "Titanium for hip and knee prostheses", in BRUNETTE D.M.T., TENGVALL P., TEXTOR M., *et al.*, (eds.), *Titanium in Medicine*, Springer, Berlin, pp. 703–746, 2001.

[LEE 04] LEE S.B., KOEPSEL R.R., MORLEY S.W., *et al.*, "Permanent nonleaching antibacterial surfaces 1. Synthesis by atom transfer radical polymerization", *Biomacromolecules*, vol. 5, pp. 877–882, 2004.

[LEG 07] LE GUEHENNEC L., SOUEIDAN A., LAYROLLE P., *et al.*, "Surface treatments of titanium dental implants for rapid osseointegration", *Dental Mater.*, vol. 23, pp. 844–854, 2007.

[LI 11] LI H., GE Y., WU Y., *et al.*, "Hydroxyapatite coating enhances polyethylene terephthalate artificial ligament graft osseointegration in the bone tunnel", *Inter. Orthop.* (SICOT), vol. 35, pp. 1561–1567, 2011.

[LI 12] LI H., YAO Z., JIANG J., *et al.*, "Biologic failure of a ligament advanced reinforcement system artificial ligament in anterior cruciate ligament reconstruction: a report of serious knee synovitis", *Arthroscopy: J. Arthrosc. Rel. Surg.*, vol. 28, pp. 583–586, 2012.

[MIC 10] MICHIARDI A., HELARY G., NGUYEN P.-C.T., *et al.*, "Bioactive polymer grafting onto titanium alloy surfaces", *Acta Biomater.*, vol. 6, pp. 667–675, 2010.

[MIN 08] MINKO S., "Grafting on solid surfaces: 'Grafting to' and 'grafting from' methods", *Polymer Surfaces and Interfaces*, Springer, Berlin, Chapter 11, pp. 215–234, 2008.

[MUL 06] MULLER R., *et al.*, "Influence of surface pretreatment of titanium and cobalt based biomaterials on covalent immobilization of fibrillar collagen", *Biomaterials*, vol. 27, pp. 4059–4068, 2006.

[OST 01] OSTUNI E., CHAPMAN R.G., LIANG M.N., *et al.*, "Self-assembled monolayers that resist the adsorption of proteins and the adhesion of bacterial and mammalian cells", *Langmuir*, vol. 17, pp. 6336–3643, 2001.

[PAL 05] PALLU S., *et al.*, "The effect of cyclo-DfKRG peptide immobilization on titanium on the adhesion and the differentiation of human osteoprogenitor cells", *Biomaterials*, vol. 26, pp. 6932–6940, 2005.

[PAR 07] PARK Y.J., SONG H.J., KIM I., *et al.*, "Surface characteristics and bioactivity of oxide film on titanium metal formed by thermal oxidation", *J. Mater. Sci. Mater. Med.*, vol. 18, pp. 565–575, 2007.

[PAV 07] PAVON-DJAVID G., GAMBLE L.J., CIOBANU M., *et al.*, "Bioactive poly (ethylene terephthalate) fibers and fabrics: grafting, chemical characterization, and biological assessment", *Biomacromolecules*, vol. 8, pp. 3317–3325, 2007.

[POR 04] PORTE-DURRIEU M.C., *et al.*, "Cyclo-(DfKRG) peptide grafting onto Ti6Al-4V: physical characterization and interest towards human osteoprogenitor cells adhesion", *Biomaterials*, vol. 25, pp. 4837–4846, 2004.

[RAM 05] RAMASUBBU N., THOMAS L.M., RAGUNATH C., *et al.*, "Structural analysis of dispersin B, a biofilm-releasing glycoside hydrolase from the periodontopathogen Actinobacillus actinomycetemcomitans", *J. Mol. Biol.*, vol. 349, pp. 475–486, 2005.

[REY 07] REYES C.D., PETRIE T.A., SCHWARTZ Z., *et al.*, "Biomolecular surface coating to enhance orthopaedic tissue healing and integration", *Biomaterials*, vol. 28, pp. 3228–3235, 2007.

[SAJ 11] SAJOVIC M., STRAHOVNIK A., DERNOVSEK M.Z., *et al.*, "Quality of life and clinical outcome comparison of semitendinosus and gracilis tendon versus patellar tendon autografts for anterior cruciate ligament reconstruction: an 11-year follow-up of a randomized controlled trial", *Am. J. Sports Med.*, vol. 39, pp. 2161–2169, 2011.

[SAL 07] SALDANA L., BARRANCO V., GONZALEZ-CARRASCO J.L., *et al.*, "Thermal oxidation enhances early interactions between human osteoblasts and alumina blasted Ti6Al4 alloy", *J. Biomed. Mater. Res.*, vol. A81, pp. 334–346, 2007.

[STA 08] STATZ A.R., BARRON A.E., MESSEERSMITH P.B., "Protein, cell and bacterial fouling resistance of polypeptoid-modified surfaces: effect of side-chain chemistry", *Soft Matter*, vol. 4, pp. 131–139, 2008.

[SUL 02] SUL Y.T., *et al.*, "Characteristics of the surface oxides on turned and electrochemically oxidized pure titanium implants up to dielectric breakdown: the oxide thickness, micropore, surface roughness, crystal structure and chemical composition", *Biomaterials*, vol. 23, pp. 491–501, 2002.

[UYA 98] UYAMA Y., KATO K., IKADA Y., "Surface modification of polymers by grafting", *Advances in Polymer Science*, vol. 137, pp. 1–39, 1998.

[VAQ 13] VAQUETTE C., VIATEAU V., GUERARD S., *et al.*, "The effect of polystyrene sodium sulfonate grafting on polyethylene terephthalate artificial ligaments on in vitro and in vivo bone tissue integration", *Biomaterials*, vol. 34, pp. 7048–7063, 2013.

[WAN 04] WANG J., LAYROLLE P., STIGER M., *et al.*, "Biomimetic and electrolytic calcium phosphate coatings on titanium alloy: physical characteristics and cell attachment", *Biomaterials*, vol. 25, pp. 583–592, 2004.

[ZAN 06] ZANTOP T., PETERSON W., SEKIYA J.K., *et al.*, "Anterior cruciate ligament anatomy and function relating to anatomical reconstruction", *Knee Surg Sports Traumatol Arthrose*, vol. 14, pp. 982–992, 2006.

[ZDY 11] ZDYRKO B., LUZINOV I., "Grafting to: polymer brushes by the 'grafting to' method", *Macromol. Rapid Commun*, vol. 32, pp. 859–869, 2011.

[ZHA 01] ZHANG F., KANG E.T., NEOH K.G., *et al.*, "Surface modification of stainless steel by grafting of poly(ethylene glycol) for reduction in protein adsorption", *Biomaterials*, vol. 22, pp. 1541–1548, 2001.

7

Biomaterial Structures for Anterior Cruciate Ligament Replacement

7.1. Introduction

The anterior cruciate ligament significantly participates in the normal knee function and stability. It is attached to the tibia and femur and transmits the forces developed during motion. Its specific anatomy and poor vascularization pattern hinder spontaneous self-regeneration once torn. This poor ability has driven the emergence of reconstructive strategies which have become an important research topic in orthopedics for many decades. Indeed, ACL is one of the most common sport injuries with a high prevalence in westernized countries. Several reconstruction strategies have been proposed and those currently employed in the clinic mostly utilize autografts. However limitations such as tissue availability, donor site morbidity and pain associated with tissue harvesting have led to the development of grafts of synthetic origin. This chapter will focus on the recent advances in artificial ligament fabrication and surface modification utilized for the manufacturing of off the shelf implants. It will further elaborate on the promise of tissue engineering for ACL replacement.

Chapter written by Cédryck VAQUETTE.

7.2. Off the shelf ligaments

7.2.1. *Non-resorbable artificial ligaments*

Artificial ligaments became tremendously popular in the early 1990s and were envisioned as the holy grail of ligament reconstruction, overcoming all limitations encountered in the autografts. This enthusiasm was abruptly stopped by the poor long-term stability which often result in catastrophic failures of these structures. Indeed, several short-to mid-term clinical retrospective studies revealed that despite appropriate initial behavior and rapid patient recovery [GLO 88], the vast majority of implanted ligaments were breaking down and had to be explanted. Most of these newly developed artificial ligaments were rated as inappropriate for clinical utilization [PAU 92]. The reasons for this immense failure are multiple and are related to the biomaterials utilized in the fabrication of the ligament which were not necessarily adequate for repetitive loading. Indeed, Paulson *et al.* reported that almost one third of the patients presented excessive knee laxicity and partial rupture of the ligament occurred in another third of the patients [PAU 92]. Guidoin *et al.* investigated the explanted failed ligaments and made some highly interesting observations [GUI 00]. They noticed that the majority of the ligaments failed at the bony interface which engendered an acute inflammatory response, a thick disorganized collagenous scar tissue surrounded the ligament preventing further tissue infiltration and hence no ligamentization could occur. In addition, this scar tissue resulted in splitting the fibers of the artificial ligaments, reducing their mechanical properties.

Most of these prosthetics were soon after removed from the market or utilized only in very challenging and specific cases. Recent improvements in the polymer quality and in fiber manufacturing have enabled the development of a better and safer polyethylene terephthalate (PET) artificial ligament known under the commercial name LARS ligament. Several recent short- and mid-term clinical studies have demonstrated the safety of this prosthetic with only a few cases of synotivitis or ruptures and LARS is now considered as a suitable prosthetic for ACL replacement [PAR 13].

PET artificial ligaments have attracted significant attention from the orthopedic community and recent advances in enhancing the performance of the ligament have focussed on the ligament biointegration. Several techniques have been developed in order to increase the bone integration of the ligament within the bone channels. This is of a high significance as slippage within the surgically created bone tunnels can occur when the ligament integration with the surrounding native bone is not properly achieved.

To this end, several surface modifications have been proposed in an attempt to increase osseointegration. This can be achieved by depositing a layer of inorganic materials onto the bony ends of the ligament as demonstrated by Li et al. who utilized a mixture of gelatine/hydroxyapatite to biofunctionalize the PET material [LI 11]. An extra-articular *in vivo* model was utilized in rabbits in order to assess the bone integration of the ligament. This revealed that the presence of a layer of hydroxyapatite, although in a discontinuous arrangement, permitted a better osseointegration, with reduced scar tissue formation and increased bone formation at the interface. This was translated in higher loads to failure in a pull out test set-up 8 weeks post-implantation. In a similar fashion, bioglass 58S particles dispersed in a gelatine solution were utilized to coat PET ligaments [LI 12]. Bioglass is a purely inorganic material whose chemical composition includes SiO_2, Na_2O, CaO and P_2O_5 in specific proportions, and which has been demonstrated to facilitate bone integration [HEN 71]. Coated and non-coated ligaments were further implanted in an extra-articular rabbit model and osseointegration was investigated at 3, 6 and 12 weeks post-implantation. This resulted in a significant reduction of the thickness of the scar tissue in the surroundings of the ligament for the coated specimens at the later time points. This also led to an increase in the biomechanical performance as measured by pull out tests at 6 and 12 weeks post-implantation. Interestingly, the levels of Bone Morphogenic Protein-2, a growth factor involved in osteogenesis, and Vascular Endothelial Growth Factor, involved in angiogenesis were up-regulated *in vivo* in the bioglass coated group. This indicates that the ionic release in the local environment from the bioglass stimulated to a greater extent bone formation and neo-vascularization, essential for bone remodeling and growth.

These promising approaches nevertheless present a number of inconveniences such as the inadequate homogeneity of the coating and the poor biomechanical stability especially during the insertion of the artificial ligament in the bone tunnels. Indeed, this can induce a significant shear stress and therefore might potentially damage and/or remove the coating. One of the other main disadvantages of the coating method is the utilization of materials from animal origin, that is, the gelatine, associated with a potential risk of disease transmission. This was rendered necessary for encapsulating and delivering the HA or bioglass particles and also permitted us to entrap the inorganic material within the artificial ligament. In a recent study, it was demonstrated that the PET can be efficiently surface modified by utilizing a technology generally employed in the semiconductor industry: pulse laser deposition (PLD). PLD was used to deposit a thin film of akernamite, a material known for its osteoindutive properties [LI 14a]. This resulted in the deposition of a continuous thin film of akernamite nanoparticles onto the PET fibers. The surface-modified ligaments were subsequently implanted in an anterior cruciate reconstruction model in a rabbit and the osseointegration was investigated at various time points up to 8 weeks post-implantation. It was observed that the akernamite coating was extremely efficient at promoting bone formation within the bone tunnels as the amount of mineralized tissue was twice that found in the control group. Similarly to the other biomaterial surface modifications previously introduced in this section, the biofunctionalization resulted in a decrease of the scar tissue formation around the implant. Another strategy for surface modification involved the grafting of polystyrene sodium sulfonate (PolyNaSS), a bioactive active polymer, for enhancing the bony integration of the artificial ligament [VAQ 13]. The bioactive polymer was covalently grafted onto the PET material ensuring a homogeneous (Figure 7.1(a)) and strong biomechanical stability of the PolyNaSS onto the ligament. The performance of the grafted ligament was studied in a pre-clinical ACL reconstruction model at 3 and 12 months and it was shown that the PolyNaSS grafting resulted in a higher frequency of bone to ligament contact at the later time point along with, here again, a reduction in the scar tissue thickness as shown in Figure 7.1(a).

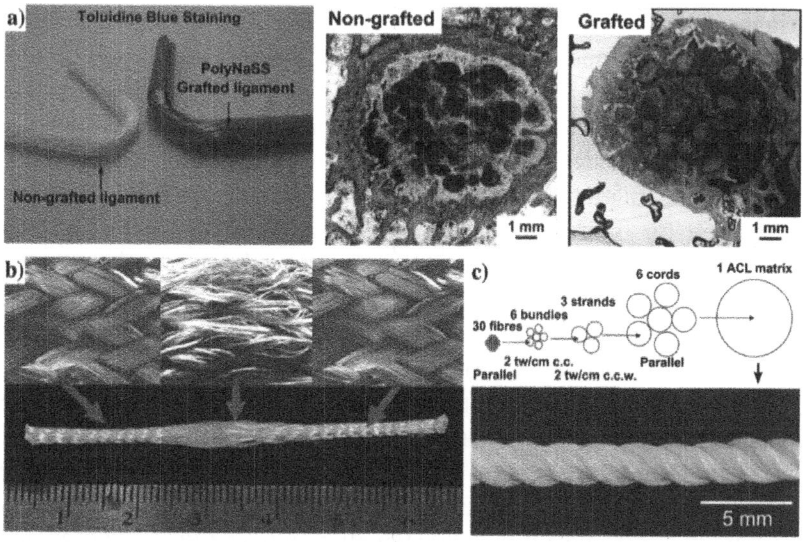

Figure 7.1. *Examples of various artificial ligaments. a) Ligament Augmentation and Reconstruction System (LARS) artificial graft. The ligament is surface modified by a bioactive polymer polystyrene sodium sulfonate (as depicted in dark gray by toluidine blue staining. This bioactivation promoted bone to ligament contact 12 months post-implantation in a preclinical ovine model [VAQ 13]. b) Structure of the PLLA artificial ligament developed by Laurencin and colleagues [LAU 05]. Similarly to most artificial implants, the ligament is composed of loosely organized fibers in the intra-articular section and is terminated by braided portions for enabling a firm attachment in the bone tunnels. c) Morphology of the silk twisted cord fabricated by Altman and colleagues [ALT 02]. The fiber twisting resulted in increased mechanical properties but unfortunately in the reduction of the pore size which can hinder cell colonization*

Although these recent advances are promising, they are not yet translated into the clinic and more research is required to fully ensure the efficacy and safety of the various surface modifications.

7.2.2. *Resorbable artificial ligaments*

One major concern in ACL reconstruction is the long-term behavior of the implanted materials, either from a biomechanical or biological point of view. Indeed, an artificial ligament may well be biomechanically adequate for supporting the physiological loading over an extended period of time; the reconstruction could fail if it triggers long-term inflammatory response. To circumvent this, scientists have attempted to develop biomaterials which will slowly degrade over time, ideally at the same pace as the ligamentization occurs. Hence, several artificial ligaments utilizing well-known biodegradable materials have emerged as potential candidates for ACL reconstruction. This is the case for the artificial ligament developed by Laurencin's group, which was fabricated by twisting and braiding various aliphatic polyester fibers (known to degrade via hydrolysis and enzymatic degradation) such as polyglycolic acid (PGA), poly-glycolic-co-lactic acid (PLGA) and poly-L-lactic acid (PLLA) into a dense and robust structure (Figure 7.1(b)) [FRE 07, LAU 05, LU 05]. These structures would initially present enough mechanical strength in order to supply the normal function of the ligament. Particular attention was paid to the degradation rate of these structures, which was an essential component of the fabrication strategy. To this end Lu *et al.* systematically tested PGA, PLGA, and PLLA braided scaffold and demonstrated that, despite initial adequate mechanical properties the PGA scaffold was degrading at a very high rate even inducing in vitro toxicity due to the excessive release of acidic degradation by-products [LU 05]. As the PLLA could maintain high mechanical properties for a long period of time, the authors claimed it was the most suitable biodegradable polymer for ACL replacement. This was consistent with the literature as PLLA has already been utilized for long-term biomedical application such as fixation screws or pins. However, PLLA has also been associated with late inflammatory response due to an enrichment in highly crystalline micro-particles with an extremely slow degradation rate [BER 95].

Another strategy involving the utilization of a very slow degradable material, that is, silk, has been implemented by Kaplan's group which were amongst the pioneers to utilize silk in biomedical applications [ALT 02]. In this approach, silk fibers extracted from *Bombyx mori* are assembled in a parallel manner and thereafter twisted into a bundle (Figure 7.1(c)). The same procedure is repeated with 6 bundles in order to create a strand and 3 strands are arranged in parallel fashion to form a cord and finally 6 parallel cords are necessary for creating the artificial ligament matrix. This multiscale organization strongly resembles the architecture and organization of the native ligament and is believed to be beneficial to the reconstruction as it increases the mechanical strength and resistance to loading. Indeed, the silk matrix displayed mechanical properties close to those of the native ligament. The advantages of this technique come from the high mechanical properties of the structures, the excellent biocompatibility of the silk, once the sericin had been removed, and from the very slow degradation pattern of the material which can potentially enable a full ligamentization before the programmed breakdown of the artificial structure. A recent study reported on the combination of a bioceramic and a silk artificial ligament [LI 14b]. In this approach, the silk ligament is inserted into a rapid prototyped tricalcium phosphate scaffold whose function is to promote osteoconduction and enhance bone to ligament contact. A pilot study in pigs involving 2 animals displayed promising results in terms of mechanical stability and tissue colonization 3 months post-surgery. However, longer pre-clinical evaluation is required to fully establish the efficacy and safety of this novel artificial ligament [LI 14b] especially 12 to 24 months post-implantation to ensure that the ligament does not experience excessive wear.

7.2.3. *Natural materials for ACL replacement*

Natural tissues from animal origin have been utilized in the past due to their excellent biocompatibility, rapid biointegration and high collagenous content which has rendered them highly attractive. This is the case for example for the Small Intestinal Submucosa better known as SIS. This tissue was popularized by Babylak's group in the mid-1990s and applied to some extent in ligament reconstruction. However the limited mechanical properties of the decellularized tissue hinder its implementation in full load-bearing application such as ACL replacement. Nevertheless, it has been used as an augmentation device for tendon regeneration [GIL 07] or rotator cuff ligament reconstruction [DEJ 01] in dogs with some success. Another drawback of the current decellularized SIS is the slow cellular infiltration once implanted. Indeed, it has been demonstrated that the dense ECM structure resulting from the decellularized process is not prone to rapid tissue colonisation. The source of SIS, decellularization and sterilization methods have also been demonstrated to impact immensely the regenerative outcomes. Indeed, the tissue organization and architecture of young tissue extracts differs in many ways to those of older animals [TOT 11]. In addition, if the decellularization process is too harsh, it can result in the destruction of the ECM structure, the removal of growth factors and proteoglycans which is in turn detrimental to the tissue regeneration [LIN 14].

Off the shelf grafts purely rely on the re-colonization of the implant by the host tissues and therefore can induce large variation in the regenerative outcome due to a difference in age and health conditions. Over the last decade, there have been attempts to include more biological components into the prosthetic used for ligament reconstruction. This has resulted in the emergence of a field called ligament tissue engineering, the paradigm of which is the combination of a biomaterial structure (generally referred to as scaffold), cells and bioactive molecules. This method is envisioned as the future of ligament reconstruction and has become a prominent area of research. The following section describes the main tissue engineering strategies developed in recent years and elaborates on the challenges that this field is facing.

7.3. Tissue-engineered constructs

7.3.1. *Cell sheet technology*

Although tissue engineering traditionally utilizes porous 3D structures for delivering the cells *in vivo*, several approaches have uniquely used cells without any carrier system. This strategy is based on the capability of the cells to form their own matrix during *in vitro* culture. This results in the formation of a so-called cell sheet, which can be physically handled without, to some extent, compromising its integrity. This is generally achieved by culturing the cells (which can be of various cell types) in 2D with a medium favorable to collagenous deposition. Hence this approach seems particularly suited for ligament or tendon tissue engineering as it recapitulates to a large extent the biological composition of the native tissue (mostly collagen type I and III). Ascorbic acid is commonly utilized at various concentrations and in combination with fibroblasts' specific growth factors for favoring the deposition of the collagenous matrix, which enables the creation of the cell sheet [MA 12]. This cell sheet remained relatively fragile but can be handled and further manipulated into a bundle resembling a tendon or a ligament (as shown in Figure 7.2(a)). A preclinical evaluation in an ACL reconstruction model in sheep demonstrated the excellent integration of these scaffold-free tissue-engineered constructs. It was further reported that the elastic modulus of the reconstructed ligament increased over 90-fold to reach 52% of that of the native contralateral ACL at 6 months post-implantation. It is worth noting that it was reported that none of the implanted ligaments failed within the time course of the pre-clinical evaluation (up to 6 months), despite the weak initial mechanical properties of the *in vitro* tissue-engineered ligament. A step further was recently performed by Ni *et al.* utilizing a similar technology and rolling up the cell sheet into a ligament-like construct (Figure 7.2(b)). A U-shaped spring was used for applying a mechanical stimulus, which in turn induced collagen fiber orientation (Figure 7.2(b)). The resulting constructs were further implanted in a rat patellar tendon window injury and were proven to significantly and positively impact the tissue regeneration [NI 13].

7.3.2. *Fibrous scaffolds*

Fibrous scaffolds received particular attention due to their similarities with the native tissue. Among the variety of constructs developed over the years for ligament tissue engineering, electrospun scaffolds have significantly contributed to the advancement of the field. Electrospinning is a manufacturing technique enabling the fabrication of non-woven nano- to micro-fibrous porous scaffolds from a solution forced through an electrostatic field. The great interest in these electrospun scaffolds comes from the possibility of aligning the polymeric fibers and hence mimicking the architectural features of the ligament. This also provides a topographical stimulus to the cells usually resulting in increased ECM synthesis and more specifically in up-regulated expressions of collagen type I and III [LEE 05] the major ECM components of tendons and ligaments. Large fibers (from hundreds of micrometers) have also been utilized for *in vitro* engineering ligamentous tissue; avian tenocytes were long-term cultured into a polyglycolic acid (PGA) non-woven scaffold [CAO 06]. The tissue-engineered constructs were further placed in a U-shape device for applying a constant deformation. This resulted in the formation of ligament-like tissue 10 weeks post-seeding, possessing a well organized and aligned collagen matrix. As the PGA degraded within the time frame of the *in vitro* culture, the resulting tissue became somehow "scaffold-free". However, the rapid degradation of the PGA could be detrimental to the mechanical properties of the construct and to the cells due to the acidic degradation products released in the media. Another approach developed by Ignatius's group involved a fibrous highly porous PLLA scaffold with a slower degradative pattern, suitable for cell attachment and infiltration [HEC 06] as shown in Figure 7.2(c). This scaffold was further utilized under uniaxial stretching in a bioreactor system for advancing the maturation of the ligament-like tissue. This resulted in an up-regulation of collagen type I and III and tenascin C, highly relevant for ligament tissue engineering [KRE 12].

7.3.3. *Knitted/braided scaffolds*

Ligament tissue engineering using a knitted scaffold was initially developed by Goh's group in Singapore. The knitted scaffold is here utilized for providing mechanical stability and strength but also for delivering mesenchymal progenitor cells to the injured tissue, which resulted in accelerated regeneration compared to the natural healing in an Achilles tendon model in rabbits [OUY 03]. The scaffold was further optimized and silk was utilized as the structural material for providing enhanced mechanical support. However, the pore size present in knitted structures does not allow an efficient cell seeding and hence a poor cell delivery is attained. To circumvent this, several strategies have been developed by incorporating a layer of materials over the macroscopic pore of these structures. This was achieved by immersing the silk scaffold into a silk solution and subsequently freeze-drying the composite construct (Figure 7.2(d)) [FAN 09]. Hydrogels such as fibrin blue [OUY 03] or alginate [VAQ 10b] can also be employed for encapsulating cells of interest as displayed in Figure 7.2(e). This resulted in better regeneration of the treated tendon in a rabbit model 12 weeks post-implantation. Another method consisted of depositing a layer of electrospun fibers whose high surface area per unit volume facilitates cell adhesion [SAH 06]. Aligned fibers can also be deposited onto the knitted structure for further increasing the deposition of relevant extracellular matrix components (Figure 7.2(f)) [VAQ 10a].

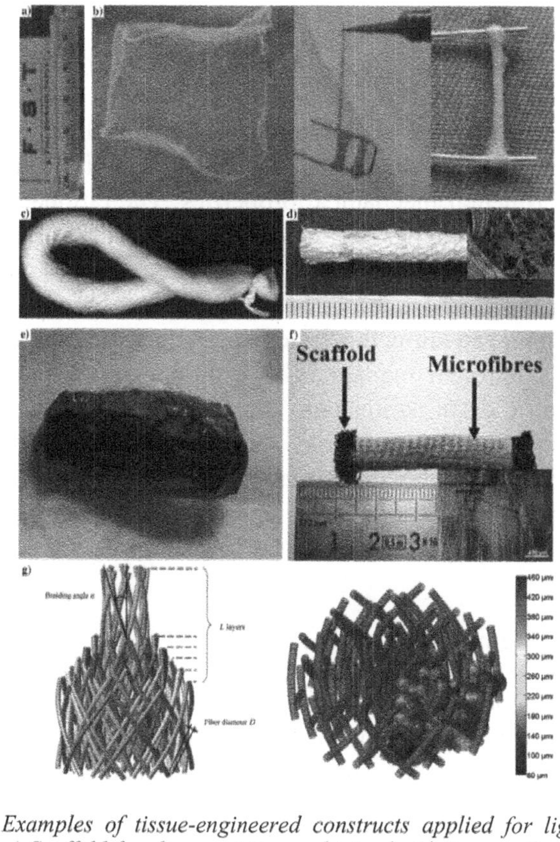

Figure 7.2. *Examples of tissue-engineered constructs applied for ligament tissue engineering. a) Scaffold-free ligament tissue obtained utilizing a cell sheet [MA 12]. b) Cell sheet technology utilized in combination with a spring for inducing matrix and cell orientation [NI 13]. c) PLLA highly porous fibrous scaffold assembled in a bundle resembling a ligament [HEC 06]. d) Knitted composite scaffold coated with a silk layer manufactured by freeze-drying. e) PLGA knitted scaffold embedded into an alginate gel for efficiently delivering stem cells [VAQ 10b]. f) Silk knitted scaffold coated with aligned electrospun polymeric fibers resulting in increasing matrix deposition [VAQ 10a]. g) Morphology of a multilayer pore-size gradient braided scaffold. The computational modeling utilized for designing the scaffold included the incorporation of interconnected pores as shown by the spheres [LAU 12, LAU 11b]*

Braided scaffolds are also suitable for ligament tissue engineering as they possess high mechanical resistance. Recently Laurent *et al.* in a series of manuscripts [LAU 11a, LAU 12, LAU 11b, LAU 14] developed a computational approach to design and fabricate a multilayered braided ligament structure as shown in Figure 7.2(g) into which a pore size gradient is created in order to enable successful and homogeneous cell and tissue infiltration. Finite element analysis considering fiber contact/friction was utilized in order to predict the geometry and the mechanical behavior of the scaffold in a large deformation framework. This method was proven highly accurate as there was an excellent match between the computational prediction and the experimental mechanical properties. The rationale behind the development of this highly sophisticated tool was to assess and control the mechanical loading and strain at the fiber level. This combined with cell seeding and subsequent culture into bioreactor applying cyclic stretching can allow the biological maturation of a neo-ligament in a highly controlled manner. This would permit us to accurately apply a specific deformation onto the fibers and hence onto the cell adhered onto the biomaterial. Biocompatibility assays have demonstrated that ovine bone marrow stem cells were capable of proliferating and infiltrating the braided structure over a 28 day period.

7.4. Concluding remarks

Tissue engineering is evolving rapidly and is a promising technology whose application is nevertheless limited by the high cost associated with the *in vitro* maturation of the neo-tissue. In addition, some the constructs developed may not be able to withstand the physiological impact once implanted in humans. Therefore, further investigations are necessary in order to translate these findings from the bench to the bedside. The other alternative, non-degradable ligaments and their recent advances, also offer some promise but significantly suffer from the reputation they gained two decades ago. Here again, long-term clinical studies are required to establish the clinical studies that are required to establish the clinical safety of these structures.

7.5. Bibliography

[ALT 02] ALTMAN G.H., HORAN R.L., LU H.H., *et al.*, "Silk matrix for tissue engineered anterior cruciate ligaments", *Biomaterials,* vol. 23, pp. 4131–4141, 2002.

[BER 95] BERGSMA J.E., DE BRUIJN W.C., ROZEMA F.R., *et al.*, "Late degradation tissue response to poly(l-lactide) bone plates and screws", *Biomaterials,* vol. 16, pp. 25–31, 1995.

[CAO 06] CAO D., LIU W., WEI X., *et al.*, "*In vitro* tendon engineering with avian tenocytes and polyglycolic acids: a preliminary report", *Tissue Engineering,* vol. 12, pp. 1369–1377, 2006.

[DEJ 01] DEJARDIN L.M., ARNOCZKY S.P., EWERS B.J., *et al.*, "Tissue-engineered rotator cuff tendon using porcine small intestine submucosa. Histologic and mechanical evaluation in dogs", *American Journal of Sports Medicine,* vol. 29, pp. 175–84, 2001.

[FAN 09] FAN H., LIU H., TOH S.L., *et al.*, "Anterior cruciate ligament regeneration using mesenchymal stem cells and silk scaffold in large animal model", *Biomaterials*, vol. 30, pp. 4967–4977, 2009.

[FRE 07] FREEMAN J.W., WOODS M.D., LAURENCIN C.T., "Tissue engineering of the anterior cruciate ligament using a braid–twist scaffold design", *Journal of Biomechanics,* vol. 40, pp. 2029–2036, 2007.

[GIL 07] GILBERT T.W., STEWART-AKERS A.M., SIMMONS-BYRD A., *et al.*, "Degradation and remodeling of small intestinal submucosa in canine Achilles tendon repair", *The Journal of Bone and Joint Surgery. American Volume*, vol. 89-A, pp. 621–630, 2007.

[GLO 88] GLOUSMAN R., SHIELDS JR C., KERLAN R., *et al.*, "Gore-Tex prosthetic ligament in anterior cruciate deficient knees", *American Journal of Sports Medicine,* vol. 16, pp. 321–326, 1988.

[GUI 00] GUIDOIN M.F., MAROIS Y., BEJUI J., *et al.*, "Analysis of retrieved polymer fiber based replacements for the ACL", *Biomaterials,* vol. 21, pp. 2461–2474, 2000.

[HEC 06] HECKMANN L., SCHLENKER H.J., FIEDLER J., *et al.*, "Human mesenchymal progenitor cell responses to a novel textured poly (L-lactide) scaffold for ligament tissue engineering", *Journal of Biomedical Materials Research – Part B Applied Biomaterials*, vol. 81, pp. 82–90, 2006.

[HEN 71] HENCH L.L., SPLINTER R.J., ALLEN W.C., *et al.*, "Bonding mechanisms at the interface of ceramic prosthetic materials", *Journal of Biomedical Materials Research Symposium,* vol. 334, pp. 117–41, 1971.

[KRE 12] KREJA L., LIEDERT A., SCHLENKER H., *et al.*, "Effects of mechanical strain on human mesenchymal stem cells and ligament fibroblasts in a textured poly(L-lactide) scaffold for ligament tissue engineering", *Journal of Materials Science: Materials in Medicine,* vol. 23, pp. 2575–2582, 2012.

[LAU 05] LAURENCIN C.T., FREEMAN J.W., "Ligament tissue engineering: an evolutionary materials science approach", *Biomaterials*, vol. 26, pp. 7530–7536, 2005.

[LAU 11a] LAURENT C., DURVILLE D., GANGHOFFER J.F., *et al.*, "Prediction of the morphological and mechanical properties of a novel scaffold for anterior cruciate ligament tissue engineering", *Computer Methods in Biomechanics and Biomedical Engineering,* vol. 14, pp. 45–46, 2011.

[LAU 11b] LAURENT C.P., GANGHOFFER J.F., BABIN J., *et al.*, "Morphological characterization of a novel scaffold for anterior cruciate ligament tissue engineering", *Journal of Biomechanical Engineering,* vol. 133, p. 133, 2011.

[LAU 12] LAURENT C.P., DURVILLE D., MAINARD D., *et al.*, "A multilayer braided scaffold for anterior cruciate ligament: mechanical modeling at the fiber scale", *Journal of the Mechanical Behavior of Biomedical Materials,* vol. 12, pp. 184–196, 2012.

[LAU 14] LAURENT C.P., VAQUETTE C., MARTIN C., *et al.*, "Towards a tissue-engineered ligament: design and preliminary evaluation of a dedicated multi-chamber tension-torsion bioreactor", *Processes*, pp. 167–179, 2014.

[LEE 05] LEE C.H., SHIN H.J., CHO I.H., *et al.*, "Nanofiber alignment and direction of mechanical strain affect the ECM production of human ACL fibroblast", *Biomaterials,* vol. 26, pp. 1261–1270, 2005.

[LI 12] LI H., CHEN S., WU Y., *et al.*, "Enhancement of the osseointegration of a polyethylene terephthalate artificial ligament graft in a bone tunnel using 58S bioglass", *International Orthopedics (SICOT),* vol. 36, pp. 191–197, 2012.

[LI 11] LI H., GE Y., WU Y., *et al.*, "Hydroxyapatite coating enhances polyethylene terephthalate artificial ligament graft osseointegration in the bone tunnel", *International Orthopedics (SICOT)*, vol. 35, pp. 1561–1567, 2011.

[LI 14a] LI H., WU C., CHANG J., *et al.*, "Functional polyethylene terephthalate with nanometer-sized bioactive glass coatings stimulating *in vitro* and *in vivo* osseointegration for anterior cruciate ligament reconstruction", *Advanced Materials Interfaces*, p. 1400027(1–10), 2014.

[LI 14b] LI X., HE J., BIAN W., *et al.*, "A novel silk–TCP–PEEK construct for anterior cruciate ligament reconstruction: an off-the shelf alternative to a bone–tendon–bone autograft", *Biofabrication,* vol. 6, p. 015010, 2014.

[LIN 14] LIN H.-K., GODIWALLA S.Y., PALMER B., *et al.*, "Understanding roles of porcine small intestinal submucosa in urinary bladder regeneration: identification of variable regenerative characteristics of small intestinal submucosa", *Tissue Engineerin:Part B,* vol. 20, pp. 73–83, 2014.

[LU 05] LU H.H., COOPER J.A., MANUELA S., *et al.*, "Anterior cruciate ligament regeneration using braided biodegradable scaffolds: *in vitro* optimization studies", *Biomaterials,* vol. 26, pp. 4805–4816, 2005.

[MA 12] MA J., SMIETANA M.J., KOSTROMINOVA T.Y., *et al.,* "Three-dimensional engineered bone–ligament–bone constructs for anterior cruciate ligament replacement", *Tissue Engineering Part A,* vol. 18, pp. 103–116, 2012.

[NI 13] NI M., RUI F.Y., TANA Q., *et al.*, "Engineered scaffold-free tendon tissue produced by tendon-derived stem cells", *Biomaterials,* vol. 34, pp. 2024–2037, 2013.

[OUY 03] OUYANG H.W., GOH J.C.H., THAMBYAH A., *et al.*, "Knitted poly-lactide-co-glycolide scaffold loaded with bone marrow stromal cells in repair and regeneration of rabbit achilles tendon", *Tissue Engineering,* vol. 9, pp. 431–439, 2003.

[PAR 13] PARCHI P.D., GIANLUCA C., DOLFI L., *et al.*, "Anterior cruciate ligament reconstruction with LARS™ artificial ligament results at a mean follow-up of eight years", *International Orthopedics (SICOT),* vol. 37, pp. 1567–1574. 2013.

[PAU 92] PAULOS L.E., ROSENBERG T., GREWE S.R., *et al.*, "The GORE-TEX anterior cruciate ligament prosthesis: A long-term followup", *American Journal of Sports and Medicine,* vol. 20, pp. 246–252, 1992.

[SAH 06] SAHOO S., OUYANG H., JAMES C.H., *et al.*, "Characterization of a novel polymeric scaffold for potential application in tendon/ligament tissue engineering", *Tissue Engineering,* vol. 12, pp. 91–99, 2006.

[TOT 11] TOTTEY, S., JOHNSON S.A., CRAPO P.M., *et al.*, "The effect of source animal age upon extracellular matrix scaffold properties", *Biomaterials,* vol. 32, 128–136, 2011.

[VAQ 10a] VAQUETTE C., KAHN C., FROCHOT C., *et al.*, "Aligned poly (L-lactic-co-e-caprolactone) electrospun microfibers and knitted structure: a novel composite scaffold for ligament tissue engineering", *Journal of Biomedical Materials Research Part A,* vol. 94A, pp. 1270–1282, 2010.

[VAQ 10b] VAQUETTE C., SLIMANI S., KAHN C.J.F., *et al.*, "A poly (lactic-co-glycolic acid) knitted scaffold for tendon tissue engineering: an *in vitro* and *in vivo* study", *Journal of Biomaterials Science, Polymer Edition,* vol. 21, pp. 1737–1760 2010.

[VAQ 13] VAQUETTE C., VIATEAU V., GUERARD S., *et al.*, "The effect of polystyrene sodium sulphonate grafting on polyethylene terephthalate artificial ligaments on *in vitro* mineralisation and *in vivo* bone tissue integration", *Biomaterials,* vol. 34, pp. 7048–7063, 2013.

8

Animal Models for Orthopedic Applications of Tissue Engineering

Animal tests constitute a step midway between *in vitro* studies and human clinical applications. Choosing an appropriate experimental model for tissue engineering purposes is critical to allow valid conclusions to be made. This chapter discusses the different factors that must be taken into account when choosing an animal model. Relevance, objectivity and reproducibility of these models as well as the genetic and immunological status of animals, and financial and ethical factors are critical issues to address. This chapter reviews the most commonly used models in osteo-articular tissue engineering studies and proposes a comprehensive decision-making approach to select the animal model which will best answer the scientific problem to be solved. Whereas preliminary evaluation of a tissue-engineered construct (TEC)'s biocompatibility and functionality is performed in small animal models and is mostly based on simple surgical procedures, preclinical evaluations must be performed in large animal models that reflect the specific human clinical setting in which the TEC will be used.

Chapter written by Véronique VIATEAU, Adeline DECAMBRON and Mathieu MANASSERO.

8.1. Introduction

Tissue engineering is an innovative strategy defined as the use of a combination of cells, engineering materials, and suitable biochemical factors to improve or replace biological functions. There is an increasing demand for new biomaterials to replace damaged osteo-articular tissues. Orthopedic applications thus represent the most important market of tissue engineering.

Tissue engineering of bone, cartilage, ligament and tendon relies on the development of tissue-engineered constructs (TEC) which combine a scaffold on which cells and/or growth factors are loaded. TEC have to be evaluated *in vivo* in experimental animal models before being used in human clinics. Animal studies are indeed essential to closing the gap between *in vitro* experiments and human clinical studies. Many animal models have been described in the field of bone, ligament, tendon and cartilage tissue engineering studies and extensive reviews of these have been made by different authors [VIA 04b, HAS 14, CHU 10]. Because the choice of the appropriate animal model is crucial to draw definitive relevant conclusions, the factors that will condition this choice must be known and well understood. The latter implies the appropriate selection of the animal (species, gender, age and when needed, hormonal status) on which testing procedures will be performed and of the experimental design in which operative and analysis procedures must be well standardized and reproducible.

8.2. Factors involved in choosing a model

Before deciding on an animal model, the problem to be solved has to be correctly identified in order to obtain the right answer to the right question. The animal species to be used and the experimental design to be selected will thus depend upon the question asked. While model relevance is the most important factor, experimental design reproducibility and morbidity, and the objectivity of data analysis, are elements of major importance. In addition, technical and financial limits may modulate the final choice.

8.2.1. *Model relevance*

A model is relevant if experimental conditions and generated effects are linked. The experimental design must therefore include the innovative strategy to be tested as well as negative and positive controls guaranteeing valid comparisons. As far as tissue engineering in the osteo-articular field is concerned, experimental objectives fall into three different categories depending on whether the TEC is submitted to: (1) a preliminary, initial evaluation to assess its biocompatibility and biofunctionality; (2) the study of the molecular and/or cellular mechanisms involved in the healing process in order to optimize its biocompatibility or its biofunctionality or (3) is evaluated in a preclinical setting.

Biocompatibility and biofunctionality testing do not require a complex biological and mechanical environment reproducing clinical-like situations. Simple tests such as animal implantation in ectopic (sub-cutaneous [VIA 13b], intra-muscular [KRU 04, VIA 04b] or more occasionally intra-peritoneal [ASH 80, BUD 80]) and orthotopic sites allow *in vivo* evaluation of biocompatibility, biodegradability and biofunctionality. Orthotopic implantation of the TEC is performed in the tissue to be reconstructed: (1) calvarial [BOS 98, HOL 90, LIN 94, SCM 90, VIL 97], ilial [AND 99], diaphyseal [JOH 87, KIT 89, MAN 13a] or metaphyseal bone defects [PAS 96] for bone TEC testing; (2) femoral condylar cartilage defects for cartilage TEC testing [CHU 10]; (3) intra-articular cruciate ligament replacement or medial collateral ligament of the knee for ligament TEC testing; (4) rotator cuff models or Achilles tendon defects [HAS 14] for tendon TEC testing. These simple surgical procedures are mostly performed in mice, rats and rabbits for financial reasons and technical simplicity. However, they are sometimes performed in large animals such as sheep [SAL 97, VIA 13c] and goats [KRU 04] which allow larger volumes of TEC to be tested.

On the contrary, preclinical evaluations of a TEC in which biocompatibility and bio-functionality has previously been evaluated, rely on animal models simulating the clinical situation in which the TEC will be used. The larger animal species indeed allow the use of operative techniques similar to the ones used in humans and treatment of lesions of clinically-relevant volumes enhancing model relevance in a preclinical setting.

The most commonly used designs in the field of bone tissue engineering are surgically induced bone defects that are known to progress to non-healing if not replaced. The notion of a critical size defect (CSD) was first described for bone by Schmitz as "the smallest intra-osseous wound that does not heal by bone formation during the lifetime of the animal" [SCH 85]. The bone CSD must be created under conditions of optimal mechanical stability in order to guarantee that non-union results exclusively from bone loss. Because its size is species and age dependant, it must be determined each time an unpublished, new CSD is used to evaluate a TEC [VIA 04, VIA 99, TOO 85]. Preclinical studies for cartilage tissue engineering also rely on the creation of cartilage defects, most of which are performed on the femoral condyles. These defects are usually performed on mini-pigs, goats, sheep and horses in which cartilage is thicker and in which larger lesions can be performed thus enhancing clinical relevance [CHU 10]. Currently used models in the field of ligament or tendon engineering have been described in goats, sheep and mini-pigs. They are based on intra-articular replacement of the Anterior Cruciate Ligament (ACL) [VIA 13a], repair procedures of patellar, Achilles or rotator cuff tendon lesions [HAS 14]. The larger animal species allow the use of operative techniques similar to the ones used in humans, enhancing model relevance in a preclinical setting.

Experimental design and associated effects must not only be correlated in the laboratory animal but also in human surgical applications. Animal anatomy, bone healing and remodeling specifications, as well as immunological and genetic status, condition model relevance.

Anatomical features

The choice of the anatomical region in which the TEC will be tested will be different depending on its future clinical applications. Calvarias (i.e. membranous bone) will be preferred if the material is to be used in cranioplasties or iliac reconstructions while metaphyseal extremities (i.e. trabecular bone) or diaphysis (i.e. cortical bone) will be preferred if the material is to be used for filling defects or segmental losses in appendicular bones, respectively. Intra-articular locations will be preferred over extra-articular locations when the TEC is designed for ACL replacement.

Specific anatomical characteristics may condition and limit the feasibility of surgical procedures in preclinical studies especially in small animal species such as rodents and rabbits. Bones and joints shapes and sizes do indeed condition surgical technical feasibility. Major disadvantages of rodents and rabbits for bone repair studies include: (1) small-sized bones, with thin and fragile cortices, thus requiring delicate surgical technique and custom-made implants for bone fixation; (2) bone TEC stability difficult to obtain in such very small defects; (3) volumes of TC to be tested not clinically relevant which makes translation to the clinical setting difficult. Major disadvantages of rodents models for cartilage repair studies include the small size of the joint and the extreme thinness of the articular cartilage, which consists of only a few cell layers. It is thus not practical, feasible or meaningful to study the effects of surgical implants in this model.

Dogs, sheep, goats and pigs have larger bones and a knee joint anatomy similar to people, allowing the use of surgical techniques and implants designed for humans. In contrast to rodents and rabbits, they allow clinically-relevant volumes of TEC to be tested. However, there are specific-related variations which must be taken into account when designing the experimentation. For example, models developed in the sheep knee for ligament replacement must take into account anatomical specificities of the joint such as smaller trochlear width, narrower femoral intercondylar notch, higher cortico-cancellous bone stock and thicker cortices in the proximal tibial metaphysis compared to humans [OST 10, NAM 14]. These differences must be taken into account when choosing the choice of a ligament TEC and the technique of its fixation to bone. In cartilage engineering studies, the size and cartilage thickness are critical issues to address when choosing a model. Cartilage defects developed in the mini-pig are preferred for preclinical studies over models developed in dogs or goats. Indeed, joint size, weight-bearing requirements, and cartilage thickness (about 1.5 mm) more closely imitate the human condition allowing assessment of full and partial thickness cartilage defects averaging the sizes of those that are of greatest clinical interest for humans. Further, second look arthroscopic evaluation of the knee joint is also possible in this species [CHU 10].

Healing and remodeling characteristics

Tissue healing and remodeling characteristics in laboratory animal depend on many variables such as blood supply, mechanical loading and most importantly, species or age dependent variables.

Species-dependant healing modalities are especially true for bone healing. Genetic variation in bone-regenerative capacity has been observed among inbred strains of mice [LI 01]. Moreover, order along the phylogenetic scale inversely correlates with the rate of bone repair: bone healing capacity is thus higher in rodents and rabbits than in other species [SCH 85]. The type and rate of bone remodeling also differ amongst species. Whereas large animals (rabbits, cats, dogs, pigs and non-human primates show Haversian-type remodeling in cortical bone, rodents do not [BEL 00]. Cortical bone remodeling in rabbits is also twice as fast as in dogs and three times as fast as in humans [SAL 97]. These features must be taken into account as they may affect material resorption: natural coral resorption is for example slower in sheep than in pigs [GUI 89].

Immature animals have higher bone healing capacity and regeneration compared to adults. Mice and rats remain exceptions to that rule as bone growth is constant throughout life in these species [BEL 00]. The magnitude of a bone CSD is inversely related to the age of the animal and bone substitutes must be evaluated in adult animals in which closed epiphyseal plates are documented with radiographs [TOO 85]. Age-dependant variables must also be taken into account for cartilage healing. Germinal cells from the physis may indeed supply regenerating cartilage and alter the study results. If the affected joint is surrounded by open growth plates, these can interfere with the applied treatment [CHU 10]. The presence of open growth plates through advancing age thus likely increase the intrinsic healing potential of osteo-chondral defects in rodents that confound repair and regeneration studies in these models. For all these reasons, while designing an animal experiment in large animals for bone and cartilage studies, skeletal maturity of the animal should be carefully considered; results in rodents and rabbits should be considered with caution and validated in other species higher on the phylogenetic scale before attempting extrapolation to humans.

Biomechanical features

Animal models used in preclinical studies for osteo-articular tissue engineering purposes cannot truly replicate the biomechanical conditions in humans. Commonly used laboratory animals are quadrupeds and subject their bone, joints, tendons and ligaments to different directions and magnitudes of load than their human counterparts, making it difficult to replicate the situation observed clinically. Differences of mechanical loading between species are also observed and must be taken into consideration when choosing an animal model and subsequently translating results to human.

Variations observed in the mechanical properties of tissues may also be observed. Variations observed in the mechanical properties of bone arise from differences in cross-sectional geometry, relative proportions of trabecular and cortical bone, amount of mineralization, degree of porosity and Haversian remodeling [VAN 01]. Interestingly, specific biomechanical behavior thus correlates more with the particular shape and function of the bone, the size of the animal and its lifestyle than with its taxonomic position.

Bone and joint loading affect the biomechanical context of the healing process. Skeletal unloading in weight bearing bones decreases the osteoblast number, bone formation rate, bone mineral density, bone maturation and mechanical strength [KOS 97]. Because joint loading also varies amongst species, some animal species will be preferred over others for preclinical studies on cartilage or ligament replacement. For example, the location of a cartilage defect should be carefully considered to avoid early overloading. In that respect, the lateral trochlea of the femur is preferred for cartilage repair and regeneration studies in the equine model. Sheep and goats, in which the stifle is carried in a more extended posture (more similar to people), will be preferred over rabbits and dogs (in which stifle flexion in standing position is pronounced), when evaluating a new TEC for cruciate ligament replacement. Caution will also be taken when extrapolating to people results obtained from intra-capsular ligament replacement in rabbits and dogs in which the stifle lies in a more flexed position and in which a higher tibial slope is noted compared to people.

Finally, for studies in which protected weight-bearing and exercise protocol are important factors, sheep, goats, pigs and horses are less well suited.

Genetic status

Unless specifically required, genetic variations are not usually taken into consideration and experimental trials are currently performed in pure breed animals (i.e. undefined genetic homozygosity) or "mongrels" (i.e. unknown genotype). Genetic variations in bone regenerative capacity nevertheless do exist as demonstrated by Li [LI 01]. However, genetic uniformity is required for cell or tissue transplantation trials: hybrid rodents, rabbits (New Zealand White rabbit), micropigs (Yucatan Micropig) or immunocompromised inbred mice (SCID, NUDE mice) are used for this purpose. Genetic selection of laboratory animals in bone engineering can also be performed to develop specific diseases (i.e. SAM mice developing osteoporosis). Transgenic pigs have also been developed to express human regulators of complement activation indicating a possibility for transgenic work in large animals [IND 02].

Immunological status

An assessment of allografts or the healing potential of various allogenic or xenogenic cellular components may benefit from implantation in constitutive (i.e. genetic) (Table 8.1), drug, surgery or radiation therapy induced immunocompromised animals. For these reasons, immunocompromised NUDE or SCID mice are currently used. As an example, athymic mice, which have a limited cellular immune response, permit initial *in vivo* study of allogenic and xenogenic cartilage regeneration strategies. Athymic rats are also available.

Mini-pigs have also been used in organ transplantation research and, therefore, have the potential to become an important large animal model for studying the use of allograft and xenograft tissues for tissue repair. Prolonged tolerance to large musculoskeletal allografts can be induced in major histocompatibility-antigen (MHC)-matched pigs with a short course of cyclosporine. Transgenic pigs have also been developed to express human regulators of complement activation, 24 indicating a possibility for transgenic work in large animals [CHU 10].

	Immunodeficiency in T lymphocyte	Immunodeficiency in B lymphocyte	Immunodeficiency in NK cells
nude	Yes	No	No
xid	No	Yes	No
Beige-Bg	Partial	No	Yes
SCID	Yes	Yes	No
Rag1	Yes	Yes	No
NIHS-Lyst^{bg} Foxn1^{nu} Btk^{xid}	Yes	Yes	Yes

Table 8.1. *Main immunodeficient strains used in in vivo studies in mice*

Physiological characteristics

Physiological status is an important factor to consider in particular in bone tissue engineering. Bone healing is indeed partly conditioned by oestrogenic metabolism in females. Ovariectomy-dependent boneloss models have been developed to mimic human osteoporosis: Mice, rats and NHP [JER 01] are currently used for this purpose. Preliminary studies have indicated that ovariectomized ferrets [MAC 95] or sheep [BEL 00] could also be interesting models for human osteoporosis. The degree of osteopenia obtained is species dependent: osteopenia in ovariectomized monkeys is mild compared with the profound cancellous osteopenia observed in ovariectomized rats. NHP models do mimic early post-menopausal changes in skeletal biology but not the disease of post-menopausal osteoporosis.

8.2.2. *Model objectivity and reproducibility*

Tissue regeneration must not only be evaluated in animals treated with the innovating TEC but also in sham-operated animals (negative controls) and in animals treated with the material considered as the gold standard (positive controls): autologous tendinous or corticocancellous bone grafts which remain the gold standard materials in ligament and bone tissue engineering, respectively. Each group should include at least 5 exploitable animals to allow analysis of statistical significance.

Accurate follow-up and assessment of tissue regeneration through conventional radiographic, *in vivo* micro-computed tomography and MRI, allow non-destructive, longitudinal, quantitative, and three-dimensional analysis of Tissue Engineering Regenerative Medicine strategies [ANA 13]. Recent availability in small animals of non-destructive functional imaging techniques such as optical imaging, positron emission tomography (PET) and single photon emission computed tomography (SPECT) open new avenues for mechanistic evaluation of the healing process. Because each imaging method provides specific information (either morphological, cellular or biomolecular), the most appropriate tool should be selected according to the requirements of the tissue engineering study [NAM 14]. In addition to imaging follow-up, a histologic and biomechanical assessment of explanted specimens' data must be achieved and results compared to those obtained from negative and positive controls.

When possible, quantitative methods will be preferred over semi-quantative techniques. Quantitative assessment of tissue healing is possible provided that animals are of similar sizes and breeds, that bone, joints or ligament sizes as well as shapes and volumes of tissue replacement are reproducible. For example, in the field of bone tissue engineering, straight bones (metatarsus, mid-femur or tibia) which allow good reproducibility in data acquisition for histomorphometry and biomechanical assays are preferred over curved bones (radius or ulna) in which only semi-quantitative methods can be used [VIA 04b].

8.2.3. *Ethical considerations*

Experimental design should follow the rule of the "3 R" (replace, reduce and refine): (1) *in vivo* experiments should only be done when alternative techniques (cell or organ cultures) fail to solve the problem; (2) the number of animals required should be the minimum required to allow valid statistical analysis; (3) when experimenting with a new design, preliminary assays should be performed on a small number of animals and models with the lowest morbidity should be selected. Morbidity should remain less than 5% as established in small and large animal models for osteo-articular research studies [AUE 00, MAT 12].

All the animal experiments should be done by experienced and specially trained personnel licensed for animal experimentation. Protocols should be evaluated by an ethics committee and tests must be performed in accordance with the usual regulations of the country. Finally, unless contradicted, pain control should systematically be provided through the per-operative and early post-operative periods.

8.2.4. *Financial considerations*

With limited research funding, costs of animal purchase and housing are important factors. In general, cost increases proportional to animal size. Models developed in rodents and rabbits are cost-effective: animals are affordable to purchase, breed and house, the volume of material to be tested is small. A large number of animals can thus be operated on allowing initial studies at low expense. In contrast, models using medium-sized to large-sized animals are expensive and are more demanding as far as housing facilities and support staff are concerned. They are essentially used in preclinical trials to allow the proof of a concept which has emerged from preliminary studies in small laboratory animals.

8.2.5. *Technical limitations*

The choice of laboratory animal must take into consideration the ease of handling as well as feasibility of postoperative care and analysis techniques. When a large animal is needed, sheep, goats or mini-pigs are often preferred over pigs which are challenging to manipulate and confine.

The choice of the experimental design is also conditioned by individual technical skills, surgical facilities: whereas subcutaneous, intramuscular and intraperitoneal implantations in rodents or rabbits can be performed without specific surgical skills and specialized instrumentation, segmental bone defect replacement, spinal fusions, cartilage defect or ligament replacement procedures do require surgical expertise and specific equipment.

The experimental design must also take into account the physical characteristics of the biomaterial. Pastes or liquid must be injected in a closed cavity to prevent leakage (i.e. sealed metaphyseal defects) and particulate or massive materials often require additional restraints to remain under stable biomechanical conditions.

Lastly, the experimental design must consider the analyzing techniques that will be applied and ensure its compatibility with them (i.e. presence of some metallic implants preclude the use of imaging techniques such MRI).

8.3. The good model for the good question research: decision-making approach

8.3.1. *Evaluation of biocompatibility, degradation and functionality*

When evaluating a new TEC, initial studies must address issues such as biocompatibility, degradation and biofunctionality. This evaluation is initially performed through the surgical implantation of the TEC in ectopic sub-cutaneous or intra-muscular sites [VIA 04b] (Figure 8.1) and subsequently in defects created in the tissue of interest: (1) circular defects created in rat calvaria [SCH 85] (Figure 8.2), rabbit femoral condyles [FLA 99] and segmental defects created in the femur, ulna or radius of mice [GAR 08, MAN 13a], rats [WER 97] and rabbits [BOL 86, TUL 86, PER 00] for bone repair studies; (2) condylar defects created in the femur of rats or rabbits for cartilage repair studies [CHU 10]; (3) extra-articular medial collateral ligament [LI 13] or intra-articular cranial cruciate [KAD 12, BAC 13] ligament replacements in rats and rabbits for ligament repair studies and; (4) flexor tendon, Achilles tendon or rotator cuff tendons replacement in hens or rats for tendon repair studies [BEA 12].

Figure 8.1. *Sub-cutaneous implantation of TEC in mice*

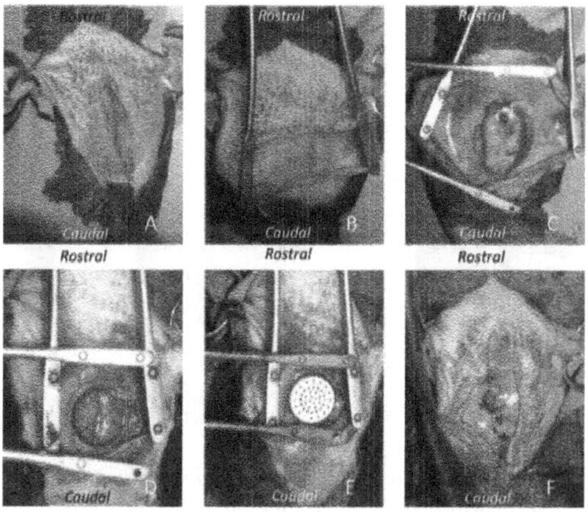

Figure 8.2. *Evaluation of TEC in a calvarial defect in a rat*

These studies are usually performed on small animal species such as mice, rats and rabbits. Small animal models like mice and rats do have several advantages: (1) expenses are low; (2) large groups of animals can be operated on; (3) homogeneity of response of strains limits individual variations of scaffold resorption and tissue formation which commonly occur in large animal models; (4) advanced imaging techniques such as 9 T MRI, microCT and bioluminescence imaging (in which MSCs labelled with luciferase can be tracked, *in vivo,* non-invasively and thus provide valuable information regarding their fate possible) are also available in these species (Figure 8.3); (5) finally, use of immune-deficient strains permits studies of human graft or cells without immune response involvement. Because many animals can be operated on, several combinations of cells/scaffolds/growth factors can be evaluated while allowing mechanistic studies of the healing process, optimization of the TEC and ultimately the selection of the TEC that will be subsequently evaluated in a preclinical setting. Animal models developed in rodents do however have some limitations: (1) they only make possible the evaluation of small, non-clinically relevant volumes of TEC which can be a major drawback in tissue engineering in which the access of the cells loaded on the scaffold to the nutrients of the recipient bed is a critical factor; (2) moreover, impact of some growth factors on healing may be species-dependant (as is the case with BMP-2 on bone formation) which generates difficulties when translating the results of preliminary studies in preclinical studies.

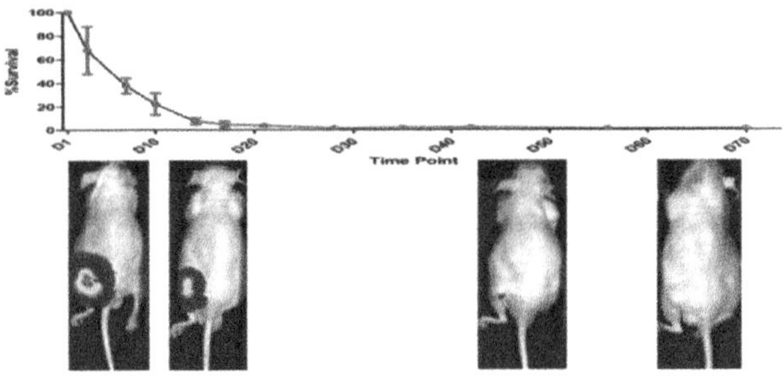

Figure 8.3. *Evaluation of transplanted mesechymeal stem cell survival by bioluminescence imaging in a TEC implanted in segmental femoral CSD bone defect in a mouse*

8.3.2. *Mechanistic studies*

Mechanistic studies include experimental designs that improve understanding of the molecular and/or cellular basis for tissue regeneration and are of the upmost importance in tissue engineering studies as they may generate new TEC designs and treatment options. These studies are currently performed in rodents in which availability of specific antibodies and of athymic, transgenic, and knockout strains makes possible the evaluation of the biological processes involved in tissue healing. Mechanistic studies require the most clinically-relevant animal models of the ones mentioned above for initial assessment of biocompatibility and biofunctionality. For example, for bone regeneration studies, a mouse or rat segmental femoral defect model will be preferred over a subcutaneous or intramuscular implantation model [MAN 13a] (see Table 8.2 and Figure 8.3).

Animal Species	Reference	Bone	Defect Length *(mm)*	Osteosynthesis	Study length *(weeks)*
Mice	[KAN 08]	Femur	5	CMP	4
	[SRO 11]	Femur	2	External fixation	8
	[GAR 08]	Femur	2	CMP locked	10
	[GAR 11]	Femur	1.8	CMP	15
	[MAN 13a]	Femur	3.5	Plate	10
Rats	[WER 97]	Femur	5	Plate	12
	[YA 92]	Femur	5	Plate *(polyethylene)*	
	[DRO 08]	Femur	6	Plate	
	[EIN 84] [EIN 99]	Femur	6	External fixator	
	[WOL 94]	Femur	8	Plate *(polyethylene)*/ pins/cerclage	
	[OES 07]	Femur	8	Plate	
	[END 06]	Tibia	6	Screw/PMMA	6
	[OZT 05]	Radius	10	O	8
Rabbits	[COO 95]	Ulna	15	0	12
	[BOL 86]	Ulna	20	0	12
	[KAR 02]	Ulna	15	0	12
	[WHE 98]	Radius	20	0	8
	[NIE 09]	Radius	15	0	16
	[GEI 05]	Radius	15	0	12
	[WIT 83]	Radius	12	0	40

Table 8.2. *Commonly used animal models for evaluation of TEC for long bone segmental replacement. CMP: centromedullary pinning*

8.3.3. *Proof of concept*

In contrast to preliminary evaluation and mechanistic studies, preclinical evaluation studies must be performed in an animal model that reflects the specific human clinical setting in which it will be used. For this reason, large animals are preferred over small animals because they allow the implantation of clinically-relevant volumes of TEC, under clinically-relevant load bearing conditions and the use of fixation implants used in people.

Bone tissue engineering

TEC designed for filling bone defects are usually evaluated in critical size bone defects created in the sheep's distal femoral metaphysic [VIA 04b]. However, such filling defects have also been described in the iliac bone [KRU 04] or spinal transverse processes in goats [KRU 06]. TEC designed for the replacement of segmental long bone defects are often tested in critical size segmental defects created in the diaphysis of a load-bearing long bone (Table 8.3). Cats [TOO 85], dogs [BRU 98, JOH 89, JOH 96], pigs [MEI 97, SEN 86], sheep [CON 00, DEN 99, GEH 93, KIR 95, KIR 98, MUI 95, REI 10, ROZ 06, WIP 94] and non-human primates [AND 78, AND 82, JER 01] have been used for that purpose. For ethical concerns, these experiments are now seldom performed on cats, dogs or non-human primates and are mostly performed on sheep (Figure 8.4). Tibial or metatarsal bone resection in sheep are thus the most commonly used animal models for testing TEC designed for segmental bone replacement [GAO 95, MAN 13b, VIA 04b].

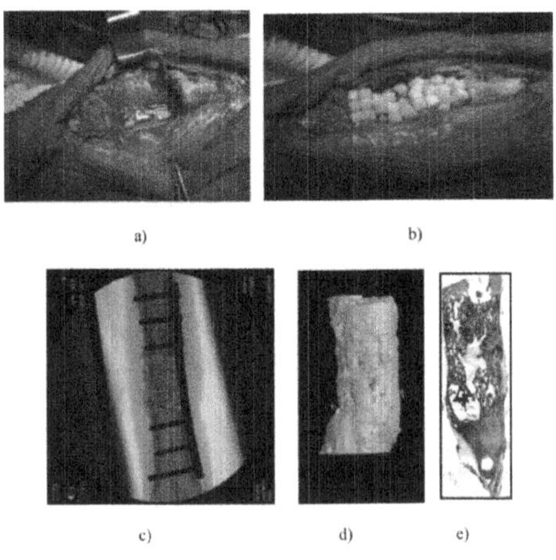

Figure 8.4. *Evaluation of TEC in a sheep metatarsal segmental bone resection. a) Empty defect, b) defect filled with the granular construct, c) postoperative radiographs, d), e) MicroCT and histology of an explanted defect, 4 months postoperatively*

Animal Species	Reference	Bone	Defect Length *(mm)*	Osteosynthesis	Study length *(weeks)*
Dogs	[BRU 98]	Femur	21	Plate	16
	[COO 95]	Ulna	25	0	12–16
Cats	[TOO 85]	Tibia	10	Plate	12
Sheep	[EHR 93]	Femur	40	External fixation	20
	[GEH 93]	Femur	25	Plate	15
	[KIR 95, KIR 98]	Femur	25	Plate	52
	[VIA 04a, MAN 13b]	Metatarsus	25	Plate	20–96
	[MUI 95]	Tibia	20	Plate	12
	[REI 10]	Tibia	30	Plate	12–60
	[ROZ 06]	Tibia	32	Plate	15
Goats	[LIU 08]	Tibia	26	External fixation	32

Table 8.3. *Commonly used large animal models for evaluation of TEC for long bone segmental replacement*

Cartilage tissue engineering

Canine, caprine, mini-pig and equine models of cartilage defect are frequently used for preclinical evaluations (Table 8.4). Although the ability to achieve critical-size defects is present for each of these models, only the equine model permits ready examination of defects at dimensions comparable to those in humans for which clinical treatment is required. Yet, many studies support the feasibility and utility of the mini-pig model for use in studying the repair and regeneration of partial thickness, full-thickness and osteo-chondral defects approaching the sizes of interest for human clinical study [CHU 10].

Ligament tissue engineering

The appropriate animal model should be chosen according to the localization – either intra-capsular or extra capsular – of the ligament to be replaced (Table 8.5). Tissue engineering studies on intra-capsular ligaments such as the ACL should be performed in cranial cruciate ligament replacement (CrCL) models developed in dogs, sheep [VIA 13a], goats or mini-pigs. Goats and sheep with increased extension of the knee and low tibial slopes should be preferred over dogs and pigs, to mimic as far as possible the biomechanical clinical setting in which the TEC will be used.

Animal Species	Reference	Joint	Defect Size *Width × deepness (mm)*	Study length *(weeks)*
Rabbits	[ORT 13]	stifle	3 × 5	3
Rabbits	[KIM 10]	stifle	3 × 2,5	3–6
Rabbits	[LAF 13]	stifle	1,4 × 2	3
Rabbits	[RAS 13]	tibia	3	3–6–12
Dogs	[WAL 13]	stifle	13.5 × 8.4 × 8.4	12–24
Mini-pigs	[MUE 09]	stifle	6 × 1	8
Pigs	[LI 09]	stifle	7	26
Sheeps	[KON 10]	stifle	7 × 9	26
Sheeps	[AKE 01]	stifle	5.4 × 6	26–52–78
Goats	[RAU 13]	stifle	3.5 × 3	26
Goats	[CHA 12]	stifle	26	26–52
Non-human primate	[BUC 03]	stifle	3.2 × 4	8

Table 8.4. *Commonly used animal models for preclinical evaluation of cartilage engineered construct*

Tendon engineering

Currently used models in the field of ligament or tendon engineering, have been described in goats, sheep and mini-pigs. Specific models have been developed according to the future fields of application of the TEC: repair procedures of lesions created in the patellar, Achilles rotator cuff tendon [HAS 14].

Animal Species	Reference	Joint	Ligament	Study length *(months)*
Rats	[LI 13]	stifle	Medial collateral	
Rats	[LUI 14]	Stifle	Cranial cruciate	
Rats		Stifle	Medial collateral	
Rabbits	[BAC 13]	Stifle	Cranial cruciate	1,5
Sheeps	[VIA 13a]	Stifle	Cranial cruciate	3 and 12
Sheeps	[KOH 13]	Stifle	Cranial cruciate	3
Dogs	[GOE 98]	Stifle	Cranial cruciate	
Pigs	[FAN 09]	Stifle	Cranial cruciate	6
Pigs	[LI 14]	Stifle	Cranial cruciate	3 and 6
Mini-pigs	[FLE 10]	Stifle	Cranial cruciate	3
Goats	[CUM 02]	Stifle	Cranial cruciate	6

Table 8.5. *Commonly used animal models for pre-clinical evaluation of TEC for ligament replacement*

8.4. Conclusions

No animal is ideal for every type of project in tissue engineering research. Because every animal model has its advantages and limitations, a comprehensive analysis of each available model needs to be conducted when planning an animal study. The research question drives the choice of the animal model. Cost effectiveness, anatomy, maturity, species-dependant healing modalities, biomechanics and relevant analysing techniques must be taken into account. Rodent models offer advantages for initial evaluation of biocompatibility, biofunctionality and mechanistic studies. To demonstrate efficacy and safety before human clinical use, long-term large animal studies evaluating clinically-relevant volumes of TEC in clinically-relevant model are needed. *In vivo* evaluation of a TEC thus involves a two-stage procedure: (1) initial evaluation of the biocompatibility, degradability and biofunctionality in a small animal model; (2) subsequent proof of concept in a clinically relevant large animal model.

8.5. Bibliography

[ANA 13] ANASTASIO M.A., LARSON J.C., BREY E.M., "Imaging challenges in biomaterials and tissue engineering", *Biomaterials*, vol. 34, no. 28, pp. 6615–6630, September 2013.

[AND 99] ANDERSON M.L., DHERT W.J., DE BRUIJN J.D., *et al.*, "Critical size defect in the goat's ilium", *Clinical Orthopaedics*, vol. 364, pp. 231–239, July 1999.

[AND 78] ANDERSSON G., GAETCHER A., GALANTE J., *et al.*, "Segmental replacement of long bones in baboons using a fiber titanium implant", *Journal of Bone and Joint Surgery*, vol. 60-A, pp. 31–40, 1978.

[AND 82] ANDERSSON G., LEREIM P., GALANTE J., *et al.*, "Segmental replacement of long bones in baboons with fiber metal implants and autologous bone grafts of different particle size", *Acta Orthopaedica Scandinavica*, vol. 53, pp. 349–354, 1982.

[ASH 80] ASHTON B.A., ALLEN T.D., HOWLET C.R., *et al.*, "Formation of bone and cartilage by marrow stromal cells in diffusion chambers *in vivo* s.c", *Clinical Orthopaedics*, vol. 151, pp. 294–307, 1980.

[AUE 00] AUER J.A., *et al.*, "Refining animal models in fracture research: seeking consensus in optimising both animal welfare and scientific validity for appropriate biomedical use", *BMC Musculoskeletal Disorders* vol. 8, p. 72, 2000.

[BAC 13] BACHY M., SHERIFI I., ZADEGAN F., *et al.*, "Anterior cruciate ligament surgery in the rabbit", *Journal of Orthopedic Surgery and Research*, vol. 8, p. 27, August 19, 2013.

[BEA 12] BEASON D.P., CONNIZZO B.K., DOURTE L.M., *et al.*, "Fiber-aligned polymer scaffolds for rotator cuff repair in a rat model", *Journal of Shoulder and Elbow Surgery*, vol. 21, no. 2, pp. 245–250, February 2012.

[BEL 00] BELLINO F.L., "Non primate models of menopause: Workshop Report", *Menopause*, vol. 7, no. 1, pp. 14–24, 2000.

[BOL 86] BOLANDER M.E., BALIAN G., "The use of demineralized bone matrix in repair of segmental defects: augmentation with extracted matrix proteins and a comparison with autologous grafts", *Journal of Bone and Joint surgery*, vol. 68-A, pp. 1264–1274 1986.

[BOS 98] BOSCH C., MELSEN B., VARGERVIK K., "Importance of the critical-size bone defect in testing bone-regenerating materials", *Journal of Craniofacial Surgery* , vol. 9, no. 4, pp. 310–316, 1998.

[BRU 98] BRUDER S.P., KRAUS K.H., GOLBERG V.M., "The effect of implants loaded with autologous mesenchymal stem cells on the healing of canine segmental bone defects", *Journal of Bone and Joint Surgery*, vol. 80-A, no. 7, pp. 985–996, 1998.

[BUC 03] BUCKWALTER J.A., MARTIN J.A., OLMSTEAD M., *et al.*, "Osteochondral repair of primate knee femoral and patellar articular surfaces: implications for preventing post-traumatic osteoarthritis", *Iowa Orthopedic Journal*, vol. 23, pp. 66–74, 2003.

[BUD 80] BUDENZ R., BERNARD G., "Osteogenesis and leukopoiesis within diffusion chamber implants of isolated bone marrow subpopulation", *American Journal of Anatomy*, vol. 159, pp. 455–474, 1980.

[CHU 10] CHU C.R., SZCZODRY M., BRUNO S., "Animal models for cartilage regeneration and repair", *Tissue Engineering Part B*, vol. 16, no. 1, pp. 105–115, 2010.

[CON 00] CONG Z., JIANXIN W., HUAIZHI F., *et al.*, "Repairing segmental bone defects with living porous ceramic cylinders: an experimental study in the dog femora", *Journal of Biomedical Materials Research*, vol. 55, pp. 28–32, 2000.

[COO 95] COOK S.D., WOLFE M.W., SALKED S.L., *et al.*, "Recombinant human osteogenic protein 1 heals segmental defects in nonhuman primates", *Journal of Bone and Joint Surgery*, vol. 77-A, pp. 734–735, 1995.

[CUM 02] CUMMINGS J.F., GROOD E.S., "The progression of anterior translation after anterior cruciate ligament reconstruction in a caprine model", *Journal of Orthopedic Research*, vol. 20, no. 5, pp. 1003–1008, September 2002.

[DEN 99] DEN BOER F.C., PATKA P., BAKKER F.C., *et al.*, "New segmental long bone defect model in sheep: quantitative analysis of healing with dual energy x-ray absorptiometry", *Journal of Orthopedic Research*, vol. 17, no. 5, pp. 654–660, 1999.

[DRO 08] DROSSE I., *et al.*, "Validation of a femoral critical size defect model for orthotopic evaluation of bone healing: a biomechanical, veterinary and trauma surgical perspective", *Tissue Engineering Part C: Methods*, vol. 14, pp. 79–88, 2008.

[EIN 99] EINHORN T.A., "Clinically applied models of bone regeneration in tissue engineering research", *Clinical Orthopaedics*, vol. 367S, pp. S59–S67, 1999.

[EIN 84] EINHORN T.A., LANE J.M., BURSTEIN A.H., *et al.*, "The healing of segmental bone defects induced by demineralized bone matrix", *Journal of Bone and Joint Surgery*, vol. 66-A, no. 2, pp. 274–279, 1984.

[EHR 93] EHRNBERG A., DE PABLOS P.J., MARTINEZ-LOTTI G., *et al.*, "Comparison of demineralized allogeneic bone matrix grafting (the Urist procedure) and the Ilizarov procedure in large diaphyseal defects in sheep", *Journal of Orthopedic Research*, vol. 11, pp. 438–447, 1993.

[ELA 12] CHAN E.F., LIU I.L., SEMLER E.J., *et al.*, "Association of 3-dimensional cartilage and bone structure with articular cartilage properties in and adjacent to autologous osteochondral grafts after 6 and 12 months in a goat model", *Cartilage*, vol. 3, no. 3, July 1, 2012

[END 06] ENDO M., *et al.*, "Bone regeneration by modified gene-activated matrix: effectiveness in segmental tibial defects in rats", *Tissue Engineering*, vol. 12, pp. 489–497, 2006.

[FAN 09] FAN H., LIU H., TOH S.L., *et al.*, "Anterior cruciate ligament regeneration using mesenchymal stem cells and silk scaffold in large animal model", *Biomaterials*, vol. 30, no. 28, pp. 4967–4977, October, 2009.

[FLE 10] FLEMING B., MAGARIAN E.M., HARRISON S.L., *et al.*, "Collagen scaffold supplementation does not improve the functional properties of the repaired anterior cruciate ligament", *Journal of Orthopedic Research*, vol. 28, no. 6, pp. 703–709, June, 2010.

[FLA 99] FLAUTRE B., DELECOURT C., BLARY M.C., *et al.*, "Volume effect on biological properties of a calcium phosphate hydraulic cement: an experimental study in the sheep", *Bone*, vol. 25, no. 2 Suppl., pp. 35S–39S, 1999.

[GAO 95] GAO T.J., "Comparative study on potential of natural coral and tricalcium phosphate cylinders in healing a segmental diaphyseal defect in sheep", in WILSON H., GREENSPAN D., (ed.), *Bioceramics*, Ponte Vedra, vol. 8, pp. 199–204, 1995.

[GAR 08] GARCIA P., *et al.*, "Development of a reliable non-union model in mice", *Journal of Surgical Research*, vol. 147, pp. 84–91, 2008.

[GAR 11] GARCIA P., *et al.*, The Locking MouseNail – a new implant for standardized stable osteosynthesis in mice", *Journal of Surgical Research*, vol. 169, pp. 220–226, 2011.

[GEI 05] GEIGER F., *et al.*, "Vascular endothelial growth factor gene-activated matrix (VEGF165 – GAM) enhances osteogenesis and angiogenesis in large segmental bone defects", *Journal of Bone and Mineral Research*, vol. 20, pp. 2028–2035, 2005.

[GEH 93] GEHRART T.N., KIRKER-HEAD C.A., KRIZ M.J., *et al.*, "Healing segmental femoral defects in sheep using recombinant human bone morphogenetic protein", *Clinical Orthopaedics*, vol. 293, pp. 317–326, 1993.

[GOE 98] GOERTZEN M.J., BUITKAMP J., CLAHSEN H., *et al.*, "Cell survival following bone-anterior cruciate ligament-bone allograft transplantation: DNA fingerprints, segregation, and collagen morphological analysis of multiple markers in the canine model", *Archives of Orthopedic and Trauma Surgery*, vol 117, nos. 4–5, pp. 208–214, 1998.

[GUI 89] GUILLEMIN G., MEUNIER A., DALLANT P., *et al.*, "Comparison of coral resorption and bone apposition with two natural corals of different porosities", *Journal of Biomedical Materials Research*, vol. 23, no. 7, pp. 765–779, 1989.

[HAS 14] HAST M.W., ZUSKOV A., SOSLOWSKY L.J., "The role of animal models in tendon research", *Bone and Joint Research*, vol. 3, pp. 193–202, 2014.

[HOL 90] HOLLINGER J.O., KLEINSCHMIDT M.D., "The critical size defect as an experimental model to test bone repair materials", *Journal of Craniomaxillofacial Surgery*, vol. 1, pp. 60–68, January, 1990.

[IND 02] INDOVINA A., BLOCK M.S., "Comparison of three bone substitutes in canine extraction sites", *Journal of Oral and Maxillofacial Surgery*, vol. 60, pp. 53–58, 2002.

[JER 01] JEROME C.P., PETERSON P.E., "Non human primate models in skeletal research", *Bone*, vol. 29, no. 1, pp. 1–6, 2001.

[JOH 96] JOHNSON K.D., AUGUST A., SCIADINI M.F., *et al.*, "Evaluation of ground cortical autograft as a bone graft material in a new canine bilateral segmental long bone defect model", *Journal of Orthopedic Trauma*, vol. 10, no. 1, pp. 28–36, 1996.

[JOH 87] JOHNSON A.L., STEIN L.E., ROE S.C., "Evaluation of collagen as a retainer for autogenous cancellous bone used in repair of full thickness cortical bone defects", *Veterinary Surgery*, vol. 16, no. 2, pp. 146–150, 1987.

[JOH 89] JOHNSON E.E., URIST M.R., SCHMALZRIED T.P., *et al.*, "Autogeneic cancellous Bone grafts in extensive segmental ulnar defects in dogs", *Clinical Orthopaedics and Related Research*, vol. 243, pp. 254–265, June 1989.

[KAD 12] KADONISHI Y., DEIE M., TAKATA T., *et al.*, "Acceleration of tendon-bone healing in anterior cruciate ligament reconstruction using an enamel matrix derivative in a rat model", *Journal of Bone and Joint Surgery (Br.)*, vol. 94, no. 2, pp. 205–209, February, 2012.

[KIR 95] KIRKER-HEAD C.A., *et al.*, "Long-term healing of bone using recombinant human bone morphogenetic protein 2", *Clinical Orthopaedics and Related Research*, vol. 318, pp. 222–30, September 1995.

[KIR 98] KIRKER-HEAD C.A., et al., "Healing bone using recombinant human bone morphogenetic protein 2 and copolymer", *Clinical Orthopaedics and Related Research*, vol. 349, pp. 205–17, April 1998.

[KAN 08] KANCZLER J.M., *et al.*, "The effect of mesenchymal populations and vascular endothelial growth factor delivered from biodegradable polymer scaffolds on bone formation", *Biomaterials*, vol. 29, pp. 1892–1900, 2008.

[KAR 02] KARAOGLU S., *et al.*, "Experimental repair of segmental bone defects in rabbits by demineralized allograft covered by free autogenous periosteum", *Injury*, vol. 33, pp. 679–683, 2002.

[KIT 89] KITSUGI T., YAMAMURO T., KOKUBO A., "Bonding behavior of a glass ceramic containing apatite and wallatsonite in segmental replacement of the rabbit tibia under load-bearing condition", *Journal of Bone and Joint Surgery*, vol. 71-A, no. 2, pp. 264–272, 1989.

[KOH 13] KOHL S., EVANGELOPOULOS D.S., KOHLHOF H., *et al.*, "Anterior crucial ligament rupture: self-healing through dynamic intraligamentary stabilization technique", *Knee Surgery, Sports Traumatology, Arthroscopy*, vol. 21, no. 3, pp. 599–605, March, 2013.

[KOS 97] KOSTENUIK P.J., HALLORAN P., MOREY-HOLTON E.R., *et al.*, "Skeletal unloading inhibits the in vitro proliferation and differentiation of rat osteoprogenitor cells", *American Journal of Physiology*, vol. 273, pp. 1133–1139, 1997.

[KRU 04] KRUYT M.C., DHERT W.J., YUAN H., *et al.*, "Bone tissue engineering in a critical size defect compared to ectopic implantations in the goat", *Journal of Orthopedic Research*, vol. 22, no. 3, pp. 544–551, May, 2004

[KRU 06] KRUYT M.C., WILSON C.E., DE BRUIJN J.D., *et al.*, "The effect of cell-based bone tissue engineering in a goat transverse process model", *Biomaterials*, vol. 27, no. 29, pp. 5099–5106, (Epub 2006 June), October, 2006

[KUR 01] KURTH A.A., MULLER R., "The effect of an osteolytic tumor on the three dimensional trabecular bone morphology in an animal model", *Skeletal Radiology*, vol. 30, pp. 94–98, 2001.

[KUS 00] KUSUMOTO K., BESSHO K., FUJIMURA K., *et al.*, "The effect of blood supply in muscle and an elevated muscle flap on endogenous tissue engineered bone by rhBMP-2 in the rat", *Annals of Plastic Surgery*, vol. 45, no. 4, pp. 408–414, 2000.

[LAF 13] LAFANTAISIE-FAVREAU C., GUZMÁN-MORALES J., SUN J., *et al.*, "Subchondral pre-solidified chitosan/blood implants elicit reproducible early osteochondral wound-repair responses including neutrophil and stromal cell chemotaxis, bone resorption and repair, enhanced repair tissue integration and delayed matrix deposition", *BMC Musculoskeletal Disorders*, vol. 14, p. 27, 2013.

[LI 09] LI W.J., CHIANG H., KUO T., *et al.*, "Evaluation of articular cartilage repair using biodegradable nanofibrous scaffolds in a swine model: a pilot study", *Journal of Tissue Engineering and Regenerative Medicine*, vol. 3, no. 1, pp. 1–10, January, 2009.

[LI 01] LI U., MASINDE G., HAMILTON-ULLAND M., *et al.*, "Genetic variation in bone-regenerative capacity among inbred strains of mice", *Bone*, vol. 29, no. 2, pp. 134–140, 2001.

[LI 13] LI H., JIANG J., GE Y., *et al.*, "Layer-by-layer hyaluronic acid-chitosan coating promoted new collagen ingrowth into a poly (ethylene terephthalate) artificial ligament in a rabbit medical collateral ligament (MCL) reconstruction model", *Journal of Biomaterials Science. Polymer Edition*, vol. 24, no. 4, pp. 431–446, 2013.

[LI 14] LI X., HE J., BIAN W., *et al.*, "A novel silk-based artificial ligament and tricalcium phosphate/polyether ether ketone anchor for anterior cruciate ligament reconstruction – safety and efficacy in a porcine model", *Acta Biomaterialia*, vol. 10, no. 8, pp. 3696–3704, August, 2014.

[LIN 94] LINDHOLM T.C., LINDHOLM T.S., MARTTINEN A., *et al.*, "Bovine bone morphogenetic protein (bBMP/ NCP)-induced repair of skull trephine defects in pigs", *Clinical Orthopaedics*, vol. 301, pp. 263–270, 1994.

[LIU 08] LIU G., *et al.*, "Repair of goat tibial defects with bone marrow stromal cells and betatricalcium phosphate", *Journal of Materials Science: Materials in Medicine*, vol. 19, pp. 2367–2376, 2008.

[LUI 14] LUI P.P., WONG O.T., LEE Y.W., "Application of tendon-derived stem cell sheet for the promotion of graft healing in anterior cruciate ligament reconstruction", *American Journal of Sports Medicine*, vol. 42, no. 3, pp. 681–690, March 2014.

[MAN 13a] MANASSERO M., VIATEAU V., MATTHYS R., *et al.*, "A novel murine femoral segmental critical-sized defect model stabilized by plate osteosynthesis for bone tissue engineering purposes", *Tissue Engineering Part C Methods*, vol. 19, no. 4, pp. 271–280, April 2013.

[MAN 13b] MANASSERO M., VIATEAU V., DESCHEPPER M., *et al.*, "Bone regeneration in sheep using acropora coral, a natural resorbable scaffold, and autologous mesenchymal stem cells", *Tissue Engineering Part A*, vol. 19, nos. 13–14, pp. 1554–1563, July 2013.

[MAT 12] MATHYS HISTING T., GARCIA P., HOLSTEIN J.H., *et al.*, "Small animal bone healing models: standards, tips, and pitfalls results of a consensus meeting", *Bone*, vol. 49, no. 4, pp. 591–9, October 2012.

[MEI 97] MEINIG R.P., BUESING C.M., HELM J., *et al.*, "Regeneration of diaphyseal bone defects using resorbable poly (L/DL-lactide) and poly (D-lactide) membranes in the Yucatan pig model", *Journal of Orthopedic Trauma*, vol. 11, no. 8, pp. 551–558, 1997.

[MUE 09] MUEHLEMAN C., LI J., ABE Y., *et al.*, "Effect of risedronate in a minipig cartilage defect model with allograft", *Journal of Orthopedic Research*, vol. 27, no. 3, pp. 360–365, March 2009.

[MUI 95] MUIR P., JOHNSON K., "Tibial intercalary allograft incorporation: comparison of fixation with locked intramedullary nail and dynamic compression plate", *Journal of Orthopedic Research*, vol. 13, pp. 132–137, 1995.

[NAM 14] NAM S.Y., RICLES L.M., SUGGS L.J., *et al.*, "Imaging strategies for tissue engineering applications", *Tissue Engineering Part B Reviews*, vol. 11, July 2014.

[NIE 09] NIEMEYER P., *et al.*, "Transplantation of human mesenchymal stem cells in a nonautogenous setting for bone regeneration in a rabbit critical-size defect model", *Acta Biomaterialia*, vol. 6, pp. 900–908, 2009.

[OES 07] OEST M.E., *et al.*, "Quantitative assessment of scaffold and growth factor-mediated repair of critically sized bone defects", *Journal of Orthopedic Research*, vol. 25, pp. 941–950, 2007

[ORT 13] ORTH P., ZURAKOWSKI D., ALINI M., *et al.*, "Reduction of sample size requirements by bilateral versus unilateral research designs in animal models for cartilage tissue engineering", *Tissue Engineering Part C Methods*, vol. 19, no. 11, pp. 885–891, November 2013.

[OST 10] OSTERHOFF G., LÖFFLER S., STEINKE H., *et al.*, "Comparative anatomical measurements of osseous structures in the ovine and human knee", *Knee*, vol. 18, no. 2, pp. 98–103, March 2010.

[OZT 05] OZTURK A.M., *et al.*, "Treatment of segmental bone defects in rats by the stimulation of bone marrow osteo-progenitor cells with prostaglandin E2", *International Orthopaedics*, vol. 29, pp. 73–77, 2005.

[PAS 96] PASQUIER G., FLAUTRE B., BLARY M.C., *et al.*, "Injectable percutaneous bone biomaterials: an experimental study in a rabbit model", *Journal of Materials Science: Materials in Medicine,*vol. 7, pp. 683–690, 1996.

[PER 00] PERKA C., SCHULTZ O., SPITZER R.S., *et al.*, "Segmental bone repair by tissue-engineered periosteal transplants with bioresorbable fleece and fibrin scaffolds in rabbits", *Biomaterials*, vol. 21, pp. 1145–1153, June 2000.

[RAS 13] RASTOGI A., SRIVASTAVA P., IQBAL Z., *et al.*, "Role of autologous chondrocyte transplantation in articular cartilage defects: an experimental study", *Indian Journal of Orthopaedics*, vol. 47, no. 2, pp. 129–134, March–April 2013.

[RAU 13] RAUB C.B., HSU S.C., CHAN E.F., *et al.*, "Microstructural remodeling of articular cartilage following defect repair by osteochondral autograft transfer", *Osteoarthritis Cartilage*, vol. 21, no. 6, pp. 860–868, June 2013.

[REI 10] REICHERT J.C., *et al.*, "Establishment of a preclinical ovine model for tibial segmental bone defect repair by applying bone tissue engineering strategies", *Tissue Engineering Part B: Reviews*, vol. 16, pp. 93–104, 2010.

[ROZ 06] ROZEN N., *et al.*, "Transplanted blood-derived endothelial progenitor cells (EPC) enhance bridging of sheep tibia critical size defects", *Bone*, vol. 45, pp. 918–924, 2009.

[SCH 94] SCHIMANDLE J.H., BODEN S.D., "The use of animal models to study spinal fusion", *Spine*, vol. 19, no. 17, pp. 1998–2006, 1994.

[SCH 85] SCHMITZ J.P., HOLLINGER J.O., "The critical size defect as an experimental model for craniomandibulofacial nonunions", *Clinical Orthopaedics*, vol. 205, pp. 299–308, 1985.

[SAL 97] SALMON R., DUNCAN W., "Determination of the critical size for nonhealing defects in the mandibular bone of sheep. Part 1: a pilot study", *Journal of the New Zealand Society of Periodontology*, vol. 81, pp. 6–15, 1997.

[SCM 90] SCHMITZ J.P., SCHWARTZ Z., HOLLINGER J.O., *et al.*, "Characterization of rat calvarial nonunion defects", *Acta Anatomica*, vol. 138, pp. 185–192, 1990.

[SEN 86] SENDOWSKI J.C.P., BRETEAU J.F., GRIMALDI J.F., *et al.*, Utilisation du squelette madréporaire en remplacement des greffons osseux: Services Vétérinaires des Armées, D.M., no. 307, DEF, DCSSA, AST, November 1986.

[SRO 11] SROUJI S., *et al.*, "A model for tissue engineering applications: femoral critical size defectin immunodeficient mice", *Tissue Engineering Part C Methods*, vol. 17, pp. 597–606, 2011.

[TUL 86] TULI S.M., SINGH A.D., "The osteoinductive property of decalcified bone matrix", *Journal of Bone and Joint Surgery*, vol. 60-B, no. 1, pp. 116–123, 1978.

[TOO 85] TOOMBS J.P., WALLACE L.J., BJORLING D.E., *et al.*, "Evaluation of Key's hypothesis in the feline tibia: an experimental model for augmented bone healing studies", *American Journal of Veterinary Research*, vol. 46, no. 2, pp. 513–518, 1985.

[TOO 85] TOOMBS J.P., WALLACE L.J., "Evaluation of autogeneic and allogeneic cortical chip grafting in a feline tibial nonunion model", *American Journal of Veterinary Research*, vol. 46, no. 2. pp. 519–528, 1985.

[VAN 01] VAN DER MEULEN M.H.C., JEPSEN K.J., MIKIC B., "Understanding bone strength: size isn't every thing", *Bone*, vol. 29, no. 2, pp. 101–104, 2001.

[VIA 13a] VIATEAU V., MANASSERO M., ANAGNOSTOU F., *et al.*, "Biological and biomechanical evaluation of the ligament advanced reinforcement system (LARS AC) in a sheep model of anterior cruciate ligament replacement: a 3-month and 12-month study", *Arthroscopy*, vol. 29, no. 6, pp. 1079–1088, June 2013.

[VIA 13b] VIATEAU V., MANASSERO M., SENSÉBÉ L., *et al.*, "Comparative study of the osteogenic ability of four different ceramic constructs in an ectopic large animal model", *Journal of Tissue Engineering and Regenerative Medicine*, vol. 20, June 2013.

[VIA 04a] VIATEAU V., GUILLEMIN G., YANG Y.C., *et al.*, "A technique for creating critical-size defects in the metatarsus of sheep for use in investigation of healing of long-bone defects", *American Journal of Veterinary Research*, vol. 65, no. 12, pp. 1653–1657, December 2004.

[VIA 04b] VIATEAU V., GUILLEMIN G., "Experimental animal models for tissue-engineered bone regeneration", in PETITE H., QUARTRO R. (eds.), *Engineered Bone*, Eurekah.com, 2004.

[VIL 97] VILJANEN V.V., LINDHOLM T.C., GAO T.J., *et al.*, "Low dosage of native allogeneic bone morphogenetic protein in repair of sheep calvaria defects", *International Journal of Oral and Maxillofacial Surgery*, vol. 26, pp. 389–393, 1997.

[WAL 13] WALDORFF E., ROESSLER B.J., ZACHOS T.A., *et al.*, "Preclinical evaluation of a novel implant for treatment of a full-thickness distal femoral focal cartilage defect", *Journal of Arthroplasty*, vol. 28, no. 8, p. 10, September 2013.

[WER 97] WERNTZ J.R., LANE J.M., KLEIN C., *et al.*, "The osteogenic potential of bone marrow to heal segmental long bone defects", *Proceedings of the 33rd Annual Meeting of the Orthopedic Research Society*, vol. 441, 1987.

[WHE 98] WHEELER D.L., *et al.*, "Radiomorphometry and biomechanical assessment of recombinant human bone morphogenetic protein 2 and polymer in rabbit radius ostectomy model", *Journal of Biomedical Materials Research*, vol. 43, pp. 365–373, 1998.

[WIP 94] WIPPERMANN B.W., ZWIPP H., SAEMANN T., *et al.*, "Healing of a segmental defect in the sheep tibia filled with a hydroxyapatite ceramic augmented by basic fibroblast growth factor and autologous bone marrow", *Proceedings of 40th Annual Meeting of the Orthopedic Research Society*, New Orleans, Louisiana, 1994.

[WIT 83] WITTBJER J., PALMER B., ROHLIN M., *et al.*, "Osteogenic activity in composite grafts of demineralized compact bone and marrow", *Clinical Orthopaedics*, vol. 173, pp. 229–238, 1983.

[WOL 94] WOLFF D., GOLDBERG V.M., STEVENSON S., "Histomorphometric analysis of the repair of a segmental diaphyseal defect with ceramic and titanium fibermetal implants: effects of bone marrow", *Journal of Orthopedic Research*, vol. 12. pp. 439–446, 1994.

[YA 92] YASKO A.W., *et al.*, "The healing of segmental bone defects, induced by recombinant human bone morphogenetic protein (rhBMP-2). A radiographic, histological, and biomechanical study in rats", *Journal of Bone and Joint Surgery (Am)*, vol. 74, pp. 659–670, 1992.

[ZHA 14] ZHAO C., OZASA Y., REISDORF R.L., *et al.*, "Brand award for outstanding Orthopedic research: engineering flexor tendon repair with lubricant, cells, and cytokines in a canine model", *Clinical Orthopaedics and Related Research*, vol. 7, June 2014.

9

Ceramic Materials for Dental Prostheses

9.1. The place of ceramics in modern prosthetic dentistry

In dentistry, full metallic crowns and bridges have been progressively abandoned due to their lack of esthetic and biocompatibility properties, notably in terms of corrosion processes. Nowadays, prostheses on natural teeth or on dental implants are most often bilayered structures composed of a metal or ceramic framework, which ensures the mechanical resistance of the prosthesis, and of a veneering ceramic layer, which is less resistant but gives a natural tooth appearance to the restoration (Figures 9.1 and 9.2). The first esthetic crown, composed of porcelain fused to a platinum post, was patented in 1885, while the actual porcelain-fused-to-metal (PFM) concept, which provides an adequate chemical and micromechanical bond between veneering ceramic and different types of metal alloys, emerged in the 1960s. Now, PFM systems are considered to be the gold standard in terms of survival rate [HEI 10]. However, ceramic frameworks started to hit the market in the 1980s. Thus, some all-ceramics are as old as non-precious alloys or titanium-based PFM systems, and already show a 40-year clinical background. All ceramics restorations definitively eliminate the disadvantages of the metal. The earliest ceramic materials used as frameworks were glass-based materials, but the appearance of computer-aided-design (CAD)

Chapter written by Amélie MAINJOT.

and computer-aided-manufacturing (CAM) processes has strongly influenced the development of dental ceramics since the 1990s, allowing the introduction of high-strength polycrystalline ceramics such as pure alumina and zirconia. These systems provide customized frameworks for dental prostheses by milling out of ceramic blocks. Globally, there is a clear economizing trend to replace hand-craftsmanship with industrial production, even for the veneering process, companies now promoting monolithic restorations, which are no longer veneered and are simply tinted and glazed. Today, dentists can either order prostheses from a dental technician or directly from a big company or even buy CAD-CAM systems designed for dental offices, which allow them to manufacture the prostheses themselves.

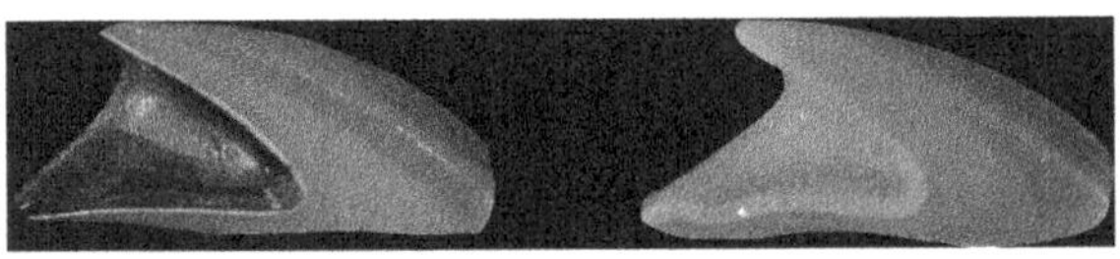

Figure 9.1. *Section through a porcelain-fused-to-metal versus an all-ceramic crown*

a)
b)
c)

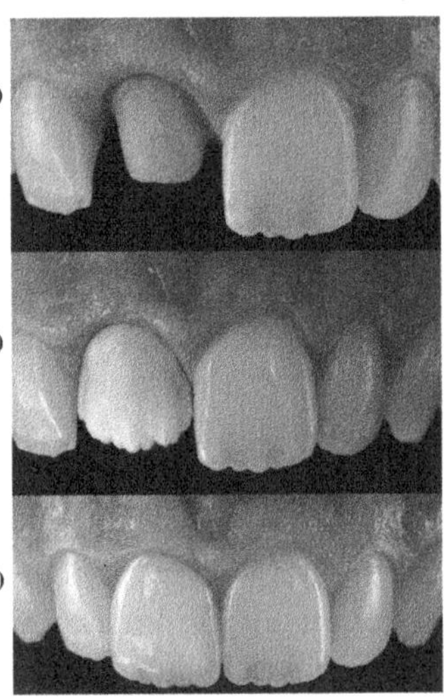

Figure 9.2. *All-ceramic crown on a central incisor: a) tooth preparation; b) zirconia framework; c) final restoration, with the veneer layer*

Dental ceramics are available in a large variety of materials in terms of chemical composition, but also in terms of manufacturing processes, which characterize a dental ceramic system. All systems possess specific properties and clinical indications with regard to their mechanical and esthetic properties, which often vary inversely. As the perfect material does not exist, dentists should be able to make, for each clinical situation, the right choice among the impressive number of products available on the market. This, however, constitutes a real challenge, especially when taking into account the rapid evolution of materials. In practice, this choice is often delegated to the dental technician, who is not able to evaluate all the clinical parameters, thus influencing the risk of failure. Indeed, ceramics are not able to be deformed as much as metal, which can convert a high stress into an irreversible deformation. Due to their fragile behavior, all ceramic restorations fail mainly due to fracture [GOO 03, PJE 07]. Cracks are initiated and propagated in ceramic materials from flaws. The risk of fracture is related to the material resistance, but also to the stress applied to the prosthesis. For example, this stress is influenced by patient habits such as bruxism (tooth grinding) [KOE 13].

Finally, due to their excellent optical properties and their chemical and physical inertias, which engender long-term stability and good biocompatibility properties, ceramics are the biomaterials of choice in modern prosthetic dentistry.

9.2. Dental ceramics systems

Dental ceramics are synthetic and inorganic materials composed of 99% oxides, with strong ionic and covalent bonds between atoms. These powders are submitted to a thermal treatment at a high temperature in order to be transformed into a dense solid (sintering process). During this process, the ceramic grains either go through the liquid phase (liquid-phase sintering), or move closer to each other remaining in a solid state (solid-phase sintering), but, in all cases, there is a retraction of the material. The sintering process can be performed before or after the prosthesis design. The type of manufacturing process is of great importance, since it can induce more or fewer flaws within the material, and thus influence the risk of crack initiation and propagation. From that point of view, industrial sintering processes used to perform CAD-CAM ceramic blocks are more efficient than hand-crafted sintering processes, which lead to a less homogeneous material and the presence of flaws.

Dental ceramics can be classified depending on their oxide chemical nature or manufacturing process or microstructure. However, the microstructural classification is the most significant from a clinical point of view, as it can be directly related to the indications and to the handling procedures of the materials. When looking at the dental ceramics microstructure, three different classes of materials can be distinguished (Figure 9.3):

– glass ceramics, which exhibit a glass matrix containing some crystals;

– infiltrated ceramics, with a matrix of crystals, containing a small amount of glass;

– polycrystalline ceramics, which contain only crystals.

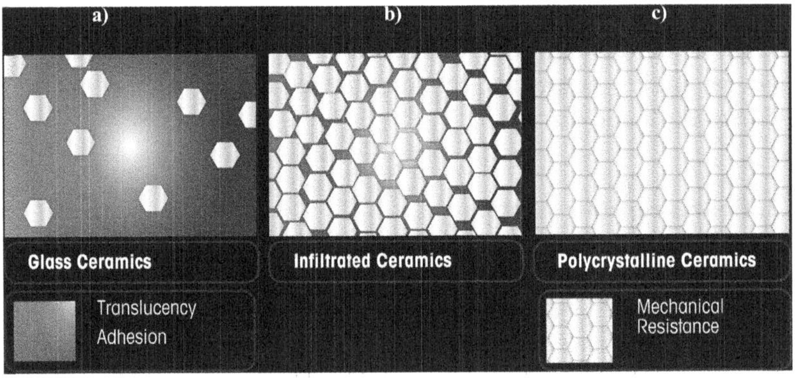

Figure 9.3. *Microstructural classification of ceramics: a) glass ceramics; b) infiltrated ceramics; c) polycrystalline ceramics*

The role of the crystals is to enhance the mechanical resistance of the material, especially because they constitute obstacles to crack propagation. The more crystals a dental ceramic contains, the more resistant it is, meaning that it can be used in high-stress areas, such as posterior regions (premolars and molars), or for larger prosthesis frameworks, which replace several teeth. Indications for each class of dental ceramics regarding mechanical resistance, and depending on the chemical nature of the crystals, are summarized in Figure 9.4. However, glass is fragile, but promotes material translucency, helping to reproduce enamel appearance, which is why veneering ceramics contain a lot of glass. Moreover, glass, which is silicium oxide in an amorphous state, can be etched with hydrofluoric acid to create the micromechanical retentions required for resin cement adhesion and then for bonding the prosthesis to tooth tissues. It can also be treated with silanes, i.e. bifunctional molecules binding to the silicium atoms, which complete adhesion mechanisms by adding a chemical bond between resin cement and ceramic. The adhesion properties are crucial from a clinical point of view, since these "bondable" materials allow the carrying out of minimally invasive treatments, for which there is no longer a need to eliminate peripheral tooth tissue to ensure the retention of the prosthesis, as was required for conical tooth preparations performed for crowns or bridges (Figure 9.2(a)). Glass-based ceramics can then be used for veneers and onlays, i.e. small-bonded prostheses, which only replace part of the tooth, on non-retentive and non-invasive preparations (Figure 9.5).

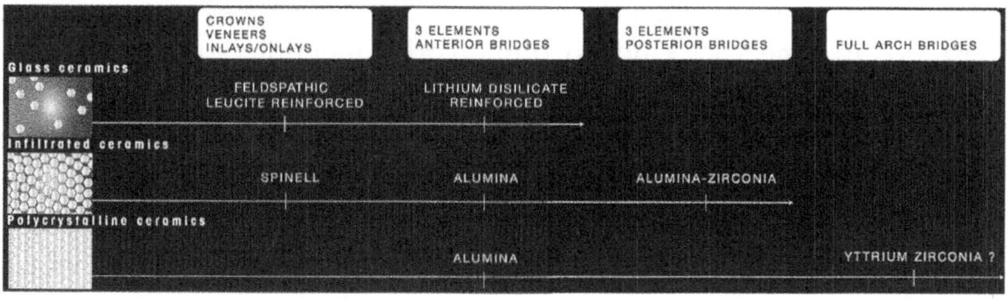

Figure 9.4. *Limits of indications of each class of dental ceramics and for each type of crystal chemical nature, respectively. From left to right, the clinical indications require more mechanical resistance*

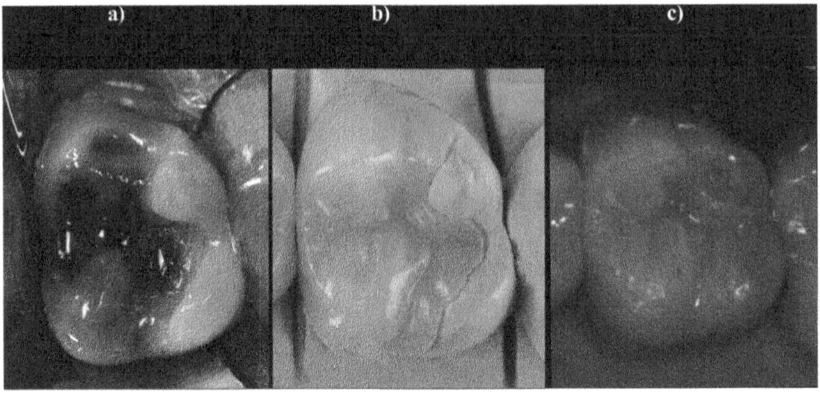

Figure 9.5. *Ceramic onlay on a molar: a) tooth preparation; b) partial restoration in lithium disilicate reinforced glass ceramic; c) bonded restoration (2-year follow-up)*

9.3. Glass ceramics

Glass ceramics constitute a large family of "bondable ceramics", which can be divided into two subclasses: classical and reinforced glass ceramics.

9.3.1. *Classical glass ceramics*

Classical glass ceramics contain a variety of feldspar crystals and a number of coloring agents. Indeed, these feldspathic ceramics are designed to veneer all kinds of frameworks, metallic or ceramic, using a hand-crafted process consisting of the application of a mix of ceramic powder and modeling liquid on the framework with a brush and the sintering in a liquid phase, layer-by-layer. They can also be used in a monolithic way to produce veneers and onlays, which are directly sintered on a replica of the restoration (Figure 9.6), or which are milled out of an industrially sintered block with a CAD-CAM system (Figure 9.7), and therefore are much more resistant. These types of restorations can be manufactured easily with a CAD-CAM system in the dental office: their long-term survival rate has been shown to be very high [OTT 08]. However, they are often monochromatic and need to be tinted on the surface to improve the esthetic result.

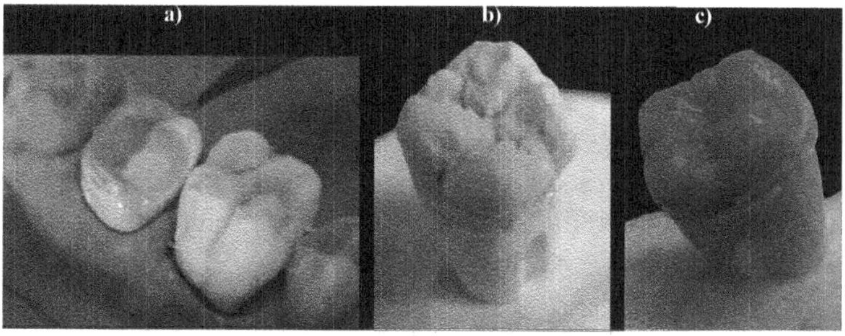

Figure 9.6. *a) Veneering ceramic artisanal stratification process, with a brush, to produce a felsdpathic onlay. The ceramic is directly sintered on an investment die replica; b) before sintering; c) after sintering*

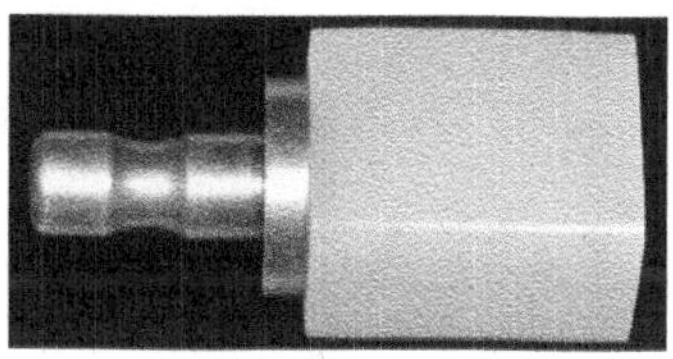

Figure 9.7. *Sintered ceramic block for a CAD-CAM manufacturing process*

9.3.2. *Reinforced glass ceramics*

Reinforced glass ceramics for dental applications started to be developed in the 1980s, in the wake of Adair and Grossmann's work [ADA 84, GRO 73], and is still experiencing growing success in single-unit prostheses. The size and the number of crystals in these materials are optimized due to specific thermal treatments done industrially, inducing and controlling crystal formation in the glass. Currently, the obtained sintered material is made in the form of a cylinder, which is softened in a special furnace by the dental technician, and then injected into a mold to create the restoration. It is also available in the form of CAD-CAM blocks for chairside or dental laboratory systems. Depending on the nature of the crystals, their mechanical properties and indications vary significantly (Figure 9.4). Due to the number, nature and elongated shape of crystals (Figure 9.8), glass ceramics reinforced with 70% by volume of lithium disilicate ($Li_2Si_2O_5$) crystals are nearly three times more resistant in terms of flexural strength than glass ceramics reinforced with 35% by volume of leucite (alumino-silicate) crystals (Figure 9.9). Thus, they can be used for a larger range of indications, particularly all kinds of single unit and bonded restorations, partial or full coverage, anterior or posterior (Figure 9.10). Restorations can be either monolithic and simply tinted, or bilayered, i.e. veneered with a feldspathic ceramic, in order to promote either mechanical or esthetic performance, respectively.

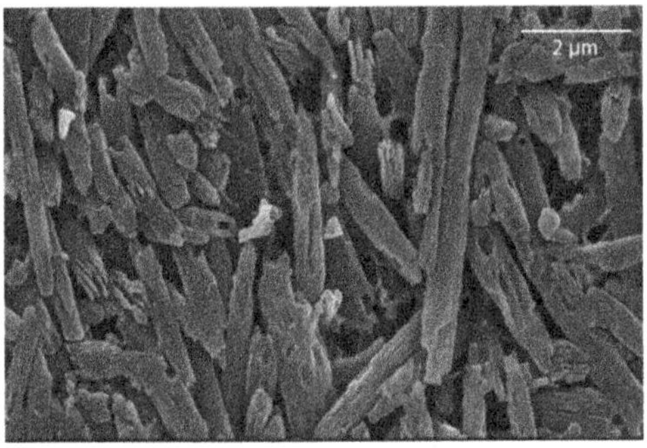

Figure 9.8. *SEM image of a lithium disilicate reinforced glass ceramic (IPS e.max Press). The glass phase has been eliminated by etching: only the characteristic elongated crystals are visible. With the courtesy of Ivoclar-Vivadent (Schaan, Liechtenstein)*

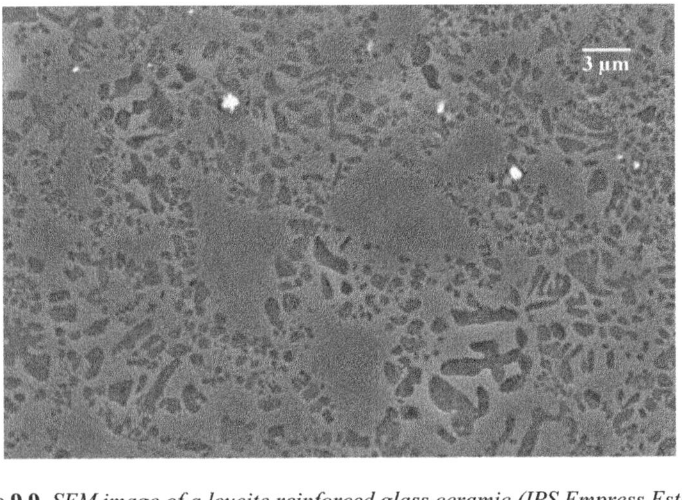

Figure 9.9. *SEM image of a leucite reinforced glass ceramic (IPS Empress Esthetic). With the courtesy of Ivoclar-Vivadent (Schaan, Liechtenstein)*

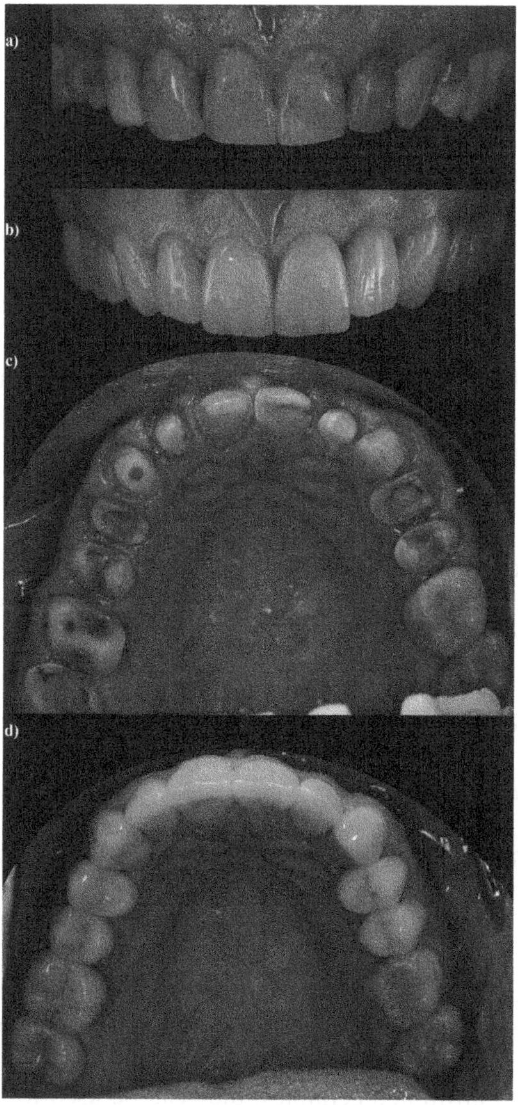

Figure 9.10. *Clinical case with lithium disilicate reinforced glass ceramic restorations (IPS e.max Press). a) Frontal view before treatment; b) frontal view after treatment; c) occlusal view of preparations. Note the large variety of the restorations to be performed: posterior partial bonded restorations, endocrowns (on premolars), veneers and crowns on natural and implant. For this typical case with single-unit restorations, lithium disilicate reinforced glass ceramic is the material of choice, since it combines the appropriate mechanical resistance and bonding capacity to resin cement and tooth structure; d) occlusal view after treatment*

9.4. Infiltrated ceramics

Infiltrated ceramics were invented in the 1980s by M. Sadoun. The crystal content reaches 74% of the material by volume and infiltrated ceramics cannot be considered as etchable because of the small amount of glass. These materials are used to produce frameworks, either by using the original hand-crafted manufacturing process called slip casting, or by the milling of CAD-CAM blocks. There are three different materials depending on the chemical nature of the crystalline matrix, with increasing mechanical resistance and opacity: spinell-based ($MgAl_2O_4$, In-Ceram Spinell, Vita Zahnfabrik, Bad Säckingen, Germany), alumina-based (Al_2O_3, In-Ceram Alumina) and a mix of alumina and zirconia (two-thirds Al_2O_3 and one-third ZrO_2, In-Ceram Zirconia). The relative opacity of these materials can be useful in masking colored tooth tissues. Despite their excellent clinical behavior and background [KER 12, SEL 13, GAL 14], infiltrated ceramics are being progressively abandoned and replaced by high-strength polycrystalline ceramics, such as zirconia.

9.5. Polycrystalline ceramics

9.5.1. *Alumina*

Polycrystalline ceramics appeared in the 1990s with CAD-CAM systems, which are able to predict and manage the significant retraction of the material occurring during the solid-phase sintering. They are not considered to be "bondable" materials, since they are not etchable, and so are not recommended to perform partial restorations. The first marketed material was alumina (Procera Alumina, Nobelbiocare, Zurich, Switzerland). Its indications are similar to alumina-based infiltrated ceramics: they can be used for prostheses of up to three elements for anterior teeth (incisors and canines) (Figure 9.3). Despite its advantages, particularly in terms of optical properties, alumina is being progressively replaced with zirconia for economic purposes.

9.5.2. *Zirconia*

Zirconia was introduced in the early 2000s and is the most recent ceramic material to be used in prosthetic dentistry. From the beginning, zirconia has been called the "white metal", the material that can replace metal in all indications, even full arch bridge frameworks of up to 14 elements, on natural teeth or implants.

Even if not ideal from an optical point of view (its refractive index is too high in comparison with natural teeth), zirconia exhibits two main advantages in comparison to other ceramic materials: its biocompatibility and mechanical properties. Indeed, zirconia shows a good cytocompatibility with osteoblasts and fibroblasts [MAN 07]. This property is very important for prostheses on implants, which pass across and are in contact with the gingival tissues (Figure 9.11). But, the most impressive characteristics of zirconia are its strength and toughness. Zirconia's flexural strength is more than two times higher than alumina and lithium dilicate reinforced glass ceramics. This is partly due to an original property: the metastable behavior of this material that engenders a toughening mechanism [CHE 09]. Indeed, pure zirconium oxide presents three crystallographic shapes depending on the temperature: cubic (c) (from 2680°C, the melting point, to 2370°C); tetragonal (t) (from 2370°C to 1170°C); and monoclinic (m) (from 1170°C to room temperature). In dental applications, zirconium oxide is not used in its pure form and is always alloyed with a dopant, which stabilizes the t phase at room temperature. This dopant is often yttrium oxide at 3 mol%. In fact, Yttria-tetragonal-zirconia-polycrystal (Y-TZP) is a temperamental material in a thermodynamically metastable state at room temperature, which enables a crystalline transformation from t to m under the effect of stress. Because a crack concentrates tensile stress on its tip, it triggers a stress-induced transformation locally, characterized by a rapid and noticeable increase in the volume of crystals (around 4%). This local volume expansion induces the development of compressive stress that closes the crack and hinders its propagation, explaining the elevated toughness of this material (Figure 9.12).

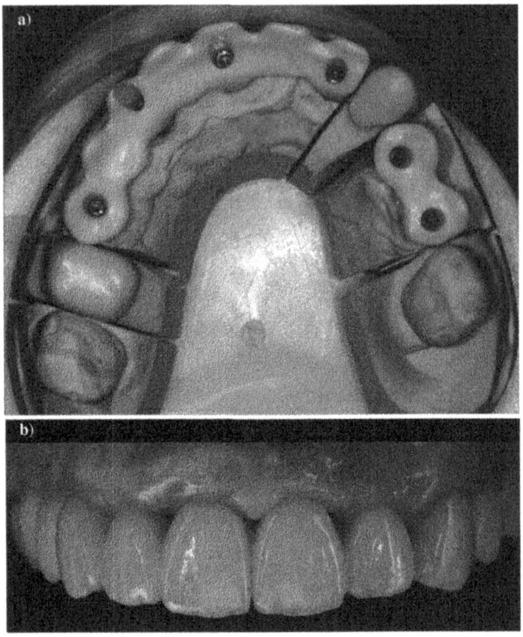

Figure 9.11. *Screwed retained zirconia bridges on implants. a) Zirconia frameworks before veneering; b) final restorations*

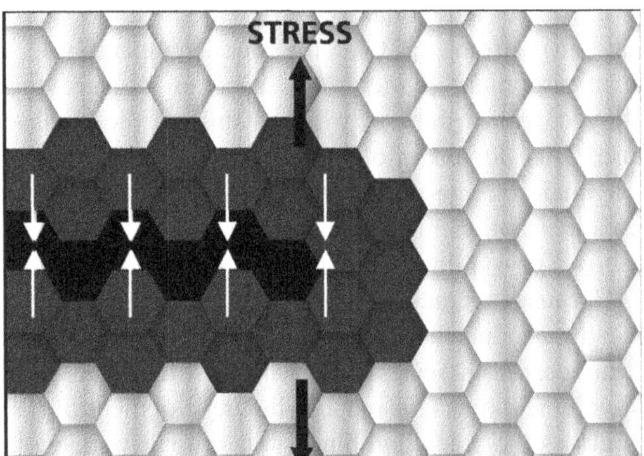

Figure 9.12. *Schematic illustration of the transformation toughening mechanism, adapted from [CHE 08]*

This metastable behavior, however, is also at the origin of an aging phenomena when placed in a moist atmosphere (low temperature degradation, LTD) (Figure 9.13). As water penetrates the crystalline structure, it generates the t-m transformation in the material surface, and subsequently engenders surface alterations, microcracks and loss of strength.

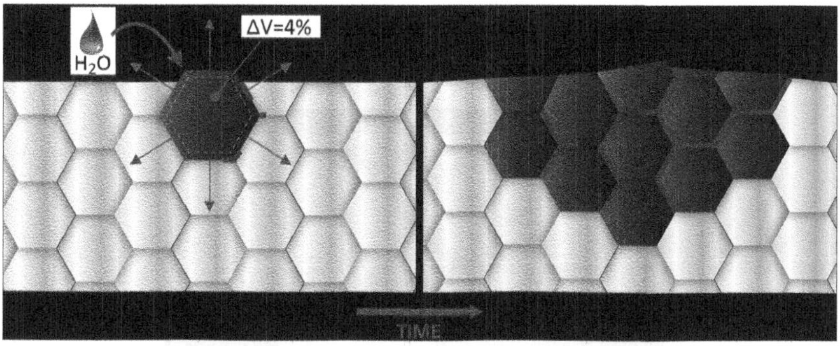

Figure 9.13. *Schematic illustration of the low temperature degradation mechanism, adapted from [CHE 08]*

Currently, the kinetics and impact of LTD on the life span of dental prostheses is still unknown [LUG 10].

Today, the clinical background of zirconia-based restorations (ZBRs) indicates high rates of short-term failures related to cohesive fractures of the veneering ceramic (around 13% after 5 years) [KOE 13] (Figure 9.14). This phenomenon, called chipping, is reported to be more frequent than with PFM restorations [HEI 10, PEL 12, VIG 12]. The chipping mechanism remains misunderstood. Several factors have been pointed out concerning this problem. Some are related to the materials and to the manufacturing process, particularly the thermal properties of zirconia and of veneering ceramic, the cooling rate, the veneer-framework thickness ratio or the design of the framework. It has also been shown that Y-TZP can undergo structural changes in the surface in contact with the veneering ceramic, due to diffusion processes or to stress development occuring during the firing procedure [DUR 12, MAI 13]. This highlights the sensitivity of this sophisticated material. Besides the material-related parameters, the clinical risk factors influencing chipping through the stress applied to the prosthesis, such as the presence of bruxism, must not be neglected [PAP 12, KOE 13]. Recent advances in ZBR include the development of translucent zirconia to perform monolithic restorations, i.e. restorations without veneering ceramic, which are simply tinted on the surface, favoring mechanical rather than optical properties. However, there is now a need to study the aging processes of these materials, which could be more metastable and more exposed to the LTD phenomenon.

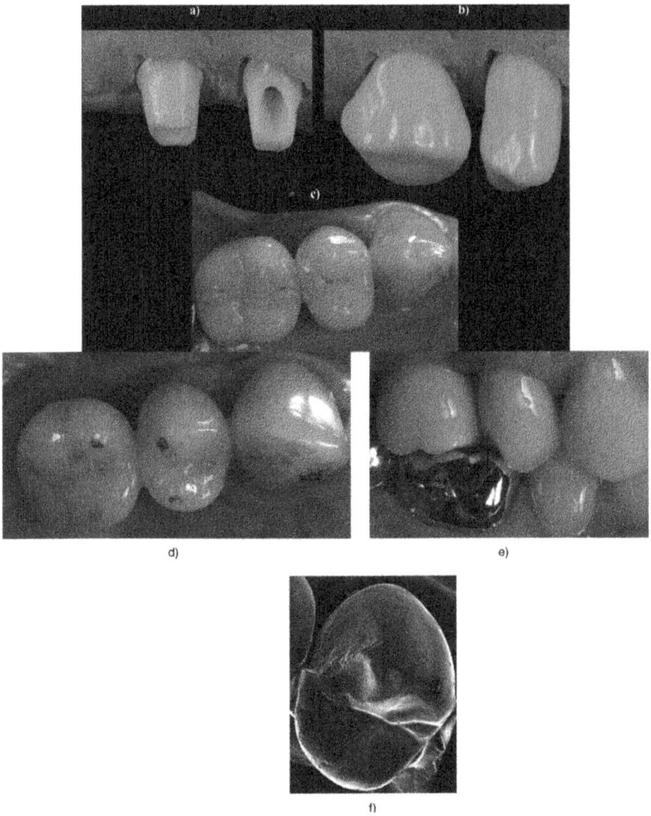

Figure 9.14. *Clinical case: zirconia crowns cemented on zirconia abutments screwed on implants. a) Zirconia abutments; b) zirconia frameworks; c) final restorations after cementation d) and e) chipping on the buccal cusp of the premolar, 3 months after placement; f) chipping by scanning-electron microscopy of the crown epoxy resin replica*

9.6. Perspectives

Ceramic materials occupy a prominent place in modern prosthetic dentistry and are progressively eliminating metal alloys from treatment options, notably due to their excellent optical properties and the absence of metallic corrosion. The choice of the type of ceramic material by the dentist is an important factor influencing the success rate of all ceramic restorations. It always constitutes a compromise between esthetics and mechanical properties and has to be oriented depending on the clinical situation and the specific indications of each ceramic family. As a constant in the history of dental biomaterials, evolution is guided by economics and now dental prosthesis manufacturing is being progressively industrialized, like everything. As shown with the emergence of monolithic restorations, the intervention of the dental technicians has been gradually reduced. However, there are still limitations to CAD-CAM technologies, whether in terms of accuracy, or environmental and economic aspects, subtractive milling processes being very costly and not ideal for the environment. In all cases, quality esthetics still require the intervention of a skillful dental technician, able to make an attractive prosthesis through the veneering process. Finally, as dental prosthesis production is confronted with globalization problems, and as many dentists and their patients are victims of economizing strategies, there is a clear need to promote quality control, keeping in mind the treatment level provided to patients.

9.7. Bibliography

[ADA 84] ADAIR P.J., Inventor of Glass ceramic dental products, 1984.

[CHE 08] CHEVALIER J., GREMILLARD L., "Zirconia ceramics and zirconia dispersed composites", in KOKUBO T.C.R.C. (ed.), *Bioceramics and Their Clinical Applications*, 2008.

[CHE 09] CHEVALIER J., GREMILLARD L., VIRKAR A.V., *et al.*, "The tetragonal-monoclinic transformation in zirconia: lessons learned and future trends", *Journal of the American Ceramic Society*, vol. 92, no. 9, pp. 1901–1920, September 2009.

[DUR 12] DURAND J.C., JACQUOT B., SALEHI H., *et al.*, "Confocal Raman microscopic analysis of the zirconia/feldspathic ceramic interface", *Dental Materials*, vol. 28, no. 6, pp. 661–671, June 2012.

[GAL 14] GALIATSATOS A.A., BERGOU D., "Clinical evaluation of anterior all-ceramic resin-bonded fixed dental prostheses", *Quintessence International*, vol. 45, no. 1, pp. 9–14, January 2014.

[GOO 03] GOODACRE C.J., BERNAL G., RUNGCHARASSAENG K., *et al.*, "Clinical complications in fixed prosthodontics", *Journal of Prosthetic Dentistry*, vol. 90, no. 1, pp. 31–41, July 2003.

[GRO 73] GROSSMAN D., Inventor of Tetrasilicic mica glassceramic method, 1973.

[HEI 10] HEINTZE S.D., ROUSSON V., "Survival of zirconia- and metal-supported fixed dental prostheses: a systematic review", *International Journal of Prosthodontics*, vol. 23, no. 6, pp. 493–502, November–December 2010.

[KER 12] KERN T., TINSCHERT J., SCHLEY J.S., *et al.*, "Five-year clinical evaluation of all-ceramic posterior FDPs made of In-Ceram Zirconia", *Int J Prosthodont*, vol. 25, no. 6, pp. 622–624, November–December 2012.

[KOE 13] KOENIG V., VANHEUSDEN A.J., LE GOFF S.O., *et al.*, "Clinical risk factors related to failures with zirconia-based restorations: an up to 9-year retrospective study", *Journal of Dentistry*, vol. 41, no. 12, pp. 1164–1174, December 2013.

[LUG 10] LUGHI V., SERGO V., "Low temperature degradation -aging- of zirconia: a critical review of the relevant aspects in dentistry", *Dent Mater*, vol. 26, no. 8, pp. 807–820, August 2010.

[MAI 13] MAINJOT A.K., DOUILLARD T., GREMILLARD L., *et al.*, "3D-characterization of the veneer-zirconia interface using FIB nano-tomography", *Dent Mater*, vol. 29, no. 2, pp. 157–165, February, 2013.

[MAN 07] MANICONE P.F., ROSSI IOMMETTI P., RAFFAELLI L., *et al.*, "Biological considerations on the use of zirconia for dental devices", *International Journal of Immunopathology and Pharmacology*, vol. 20, no. 1, Suppl 1, pp. 9–12, January–March 2007.

[OTT 08] OTTO T., SCHNEIDER D., "Long-term clinical results of chairside Cerec CAD/CAM inlays and onlays: a case series", *Int J Prosthodont*, vol. 21, no. 1, pp. 53–59, January–February 2008.

[PAP 12] PAPASPYRIDAKOS P., LAL K., "Computer-assisted design/computer-assisted manufacturing zirconia implant fixed complete prostheses: clinical results and technical complications up to 4 years of function", *Clinical Oral Implants Research*, 13 March 2012.

[PEL 12] PELAEZ J., COGOLLUDO P.G., SERRANO B., *et al.*, "A four-year prospective clinical evaluation of zirconia and metal-ceramic posterior fixed dental prostheses", *Int J Prosthodont*, vol. 25, no. 5, pp. 451–458, September–October 2012.

[PJE 07] PJETURSSON B.E., SAILER I., ZWAHLEN M., *et al.*, "A systematic review of the survival and complication rates of all-ceramic and metal-ceramic reconstructions after an observation period of at least 3 years", *Part I: Single crowns*, *Clin Oral Implants Res*, vol. 18, Suppl 3, pp. 73–85, June 2007.

[SEL 13] SELZ C.F., STRUB J.R., VACH K., *et al.*, "Long-term performance of posterior InCeram Alumina crowns cemented with different luting agents: a prospective, randomized clinical split-mouth study over 5 years", *Clinical Oral Investigations*, November 22 2013.

[VIG 12] VIGOLO P., MUTINELLI S., "Evaluation of zirconium-oxide-based ceramic single-unit posterior fixed dental prostheses (FDPs) generated with two CAD/CAM systems compared to porcelain-fused-to-metal single-unit posterior FDPs: a 5-year clinical prospective study", *Journal of Prosthodontics*, 19 February 19, 2012.

10

Dental Adhesives

10.1. Introduction

Dental adhesive systems are commonly employed to achieve a strong bond between the tooth substrate (dentin or enamel) and restorative material used. Sustained progress in the field of dental adhesive technology has provided clinicians with products and systems allowing for an increasingly conservative approach to the esthetic restoration of damaged teeth. All of them are supposedly capable of creating strong and reliable bonds to the remaining tooth tissues. These adhesive systems require the conditioning of dental tissues in order to achieve adhesion. This is done by acid etching that promotes the superficial demineralization of the substrate allowing for the replacement of minerals removed from the hard tissues by resin monomers. Upon polymerization, they become micromechanically interlocked in the created porosities [VAN 03]. However, the heterogeneity and water content of dentin extracellular matrix (ECM), infiltrated during the bonding process implies a weaker adhesion over time as compared to enamel [DE 05, BRE 08]. Indeed, while conditions for optimal bonding to enamel are now standard practice, bonding to dentin remains dependent on understanding the variations of the complex structure and properties of this substrate.

Chapter written by Mathieu DERBANNE, Stéphane Le GOFF and Jean-Pierre ATTAL.

10.2. The different adhesive systems

Every bonding system developed during the last two decades has been based on the total-etch concept (i.e. simultaneous etching of enamel and dentin) introduced by Fusayama leading to the development of the fourth-generation dental adhesive systems [FUS 80]. During past years, adhesive technology has quickly evolved from etch and rinse (ER) systems requiring separate acid etching of dental tissues followed by rinsing to more user-friendly, quicker to use self-etching systems (SESs) that have rapidly gained popularity among practitioners [VAN 11]. Contrary to ER systems, SES does not require a separate etching step. Etching is done by acidic monomers contained in the primer or in the bonding resin in the case of all-in-one systems allowing us to simultaneously condition and prime the dental substrate [PEU 05].

<table>
<tr><th></th><th>ER3</th><th>ER2</th><th>SES2</th><th>SES1</th></tr>
<tr><td>Etching</td><td>H_3PO_4
Rinsing</td><td>H_3PO_4
Rinsing</td><td rowspan="2">Acidic function bearing monomers

Hydrophilic and hydrophobic monomers

Solvent</td><td rowspan="3">Acidic function bearing monomers

Hydrophilic and hydrophobic monomers

Hydrophobic monomers

Solvent</td></tr>
<tr><td>Priming</td><td>Hydrophilic and hydrophobic monomers

Solvent</td><td rowspan="2">Hydrophilic and hydrophobic monomers

Hydrophobi c monomers

Solvent</td></tr>
<tr><td>Bonding</td><td>Hydrophobic monomers</td><td>Hydrophobic monomers</td></tr>
<tr><td></td><td>3
Steps</td><td>2
Steps</td><td>2
Steps</td><td>1
Step</td></tr>
</table>

Table 10.1. *Classification of current dental adhesive systems according to Degrange and Van Meerbeek*

Starting with the so-called fourth generation of dental adhesive systems that are in fact ER3 systems, all systems currently marketed can be classified in one of the following categories according to their mode of etching and number of steps required to achieve bonding [VAN 01, DEG 05]: ER3 and ER2 requiring separate acid etching, usually with a solution of orthophosphoric acid (H_3PO_4) and needing, respectively, three steps (etching, priming and bonding) and two steps (etching, priming and bonding are combined); SES2 and SES1 not requiring separate acid etching due to their self-etching properties and needing, respectively, two steps (self-etching priming and bonding) or one step (self-etching priming and bonding are combined in a single bottle). These categories are summarized in Table 10.1.

10.3. General principles of bonding to mineralized dental tissues

All recent developments regarding bonding to mineralized dental tissues take root in Buonocore's proposal in 1955: the use of an acidic solution (usually orthophosphoric acid) to etch the enamel surface [BUO 55]. The solution is then removed by thorough rinsing with a water spray followed by application of the adhesive system. Acid etching of dentin was introduced by Fusayama in 1979 in order to improve bonding to dental tissues, hence the "total-etch concept" that requires simultaneous acid etching of enamel and dentin as proposed by Bertolotti in 1991 [FUS 79, BER 91]. This concept still prevails today.

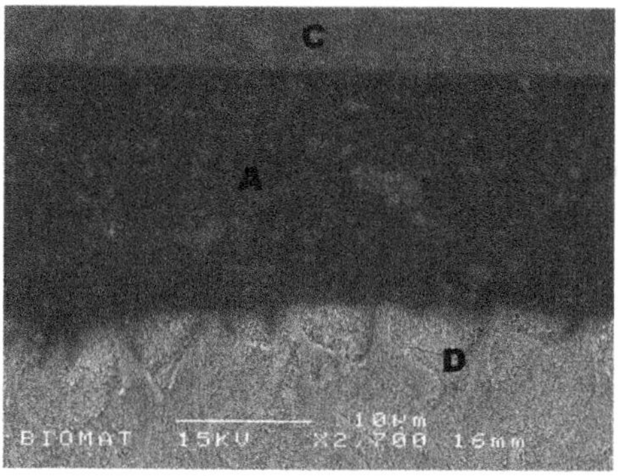

Figure 10.1. *Bonded interface on human enamel (C: composite, A: adhesive (resin) and D: enamel)*

Bonding to enamel is mainly caused by micromechanical interlocking (Figure 10.1) between the etched enamel surface and bonding material. Acid etching allows for a selective dissolution of the topmost enamel layers, developing the exposed surface and creating the condition for micromechanical retention as illustrated in Figure 10.2 [BUO 67].

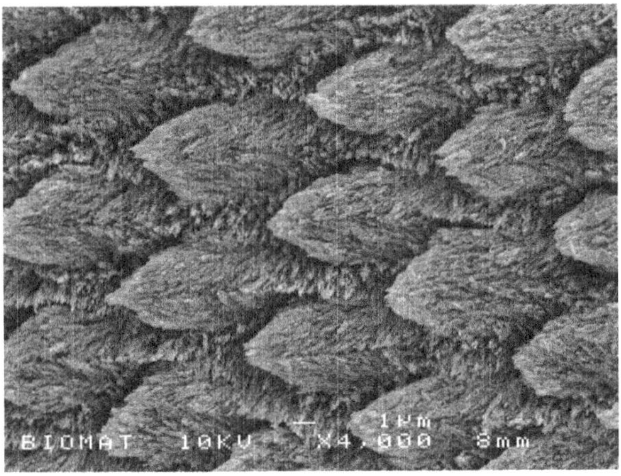

Figure 10.2. *Typical etching pattern on human enamel*

With regard to dentin, acid etching leads to superficial demineralization, exposing the embedded collagen network of dentin (Figure 10.3). This network, free of its mineral content, is kept open by water forming a 5–8 μm sponge-like structure [PAS 11]. Penetration of the bonding material in this structure, displacing water, allows for the creation of a hybrid layer composed of collagen and resin. The process, described by Nakabayashi in 1998, is usually referred to as hybridization [NAK 98].

The exposed collagen layer has by nature a very hydrophilic function rendering direct infiltration of a hydrophobic resin nearly impossible. Water needs to be displaced first. This is done through the use of a low viscosity priming solution composed of low molecular weight hydrophilic function bearing monomers and ethanol or acetone-based solvent. Infiltration of this priming solution in the exposed collagen layer allows for subsequent infiltration of a hydrophobic bonding resin, the polymerization of which will lead to hybridization (Figure 10.4) [PAS 11].

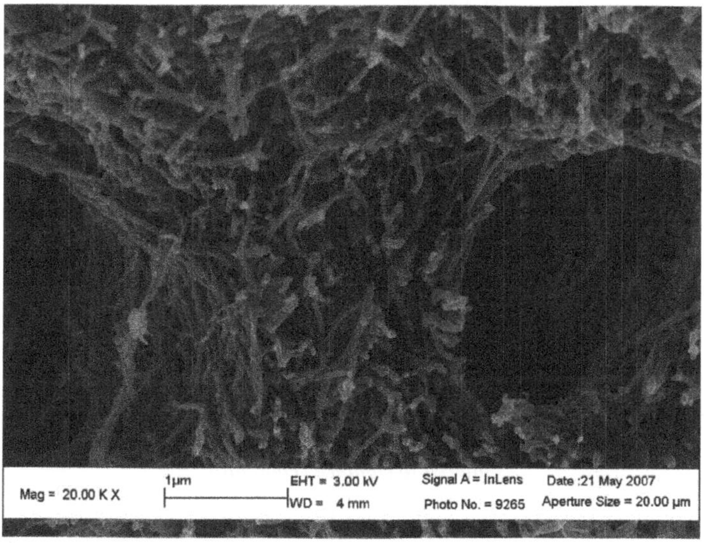

Figure 10.3. *Etched human dentin showing the embedded collagen network*

More recently, an alternative to acid etching with separate acidic solution was introduced in the form of self-etching dental adhesive systems (SES). They still, however, are capable of acid etching through the use of acidic function bearing monomers allowing for simultaneous demineralization and infiltration of dental substrates. Moreover, besides their higher user-friendliness, they allow for the conservation of a dentinal smear layer, a surface layer composed of mineral, collagenic and bacterial residues produced during the preparation of the dental cavity. Contrary to the orthophosphoric acid solution that is able to totally dissolve this layer, the less acidic SES only dissolves it partially and infiltrates it. As a result, dentinal permeability is lessened, leading to less frequent postoperative sensitivity [TAY 02, PER 03, UNE 04, PEU 05, AKI 07, VAN 07, PEU 10].

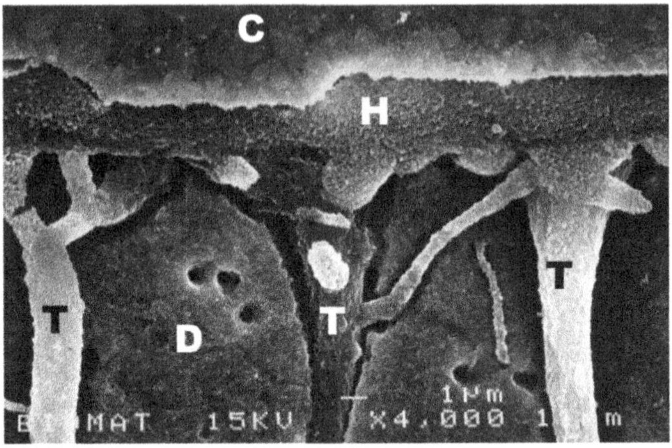

Figure 10.4. *Side view of hybrid layer after cryofracturation (C: composite, H: hybrid layer, T: resin tags inside dentinal tubules and D: dentin). Image has been digitally altered to enhance readability*

While classically described as a pure mechanical phenomenon, recent studies have shown that bonding to mineralized dental tissues also implies a physicochemical mechanism. It has been shown that reactive functions of some monomers have the ability to interact with the mineral phase of dental tissues, forming primary chemical liaisons. This phenomenon adds to micromechanical retention, improving overall adhesion [YOS 00, YOS 04].

10.4. A word on dental bonding system composition

With very few exceptions, current dental adhesive systems are all based on methacrylic monomers that allow for *in-situ* polymerization through the use of a photo-initiating system and subsequent copolymerization with the restorative material. A summarized composition of dental bonding systems is presented in Table 10.2 while common components are shown in Figure 10.5 [VAN 07].

	ER	SES
Etching	H_3PO_4 ≈ 35% in aqueous gel solution 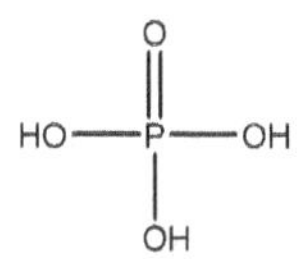	Solvent (water + cosolvent) Acidic monomethacrylates (initiator + inhibitor)
Priming	Solvent Monomethacrylates (initiator + inhibitor)	
Bonding	Dimethacrylates HEMA Initiator + inhibitor (Filler)	Dimethacrylates (HEMA) Initiator + inhibitor (Filler)

Table 10.2. *Summarized composition of current dental bonding systems (components inside parentheses are not always present)*

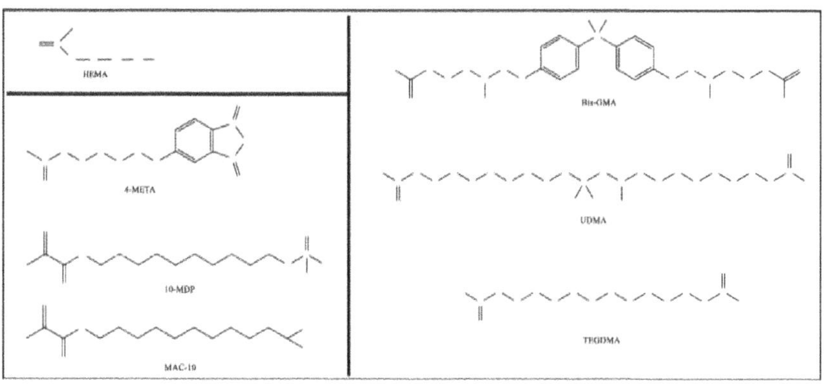

Figure 10.5. *Examples of chemical components found in dental bonding systems (upper left: HEMA, lower left: functional monomers, right: crosslinking monomers)*

While the camphorquinone (CQ)/coinitiator (usually, a tertiary amine) is the most frequently used photo-initiating system used in dental adhesive, it is sometimes complemented by other initiating systems that act synergistically [PAR 99]. Among those are 1-phenyl-1, 2 propanedione (PPD) or acylphosphine oxydes. Some adhesives are designed to be chemically polymerized when used in dark conditions (example: root canal), they contain chemical initiating systems like benzoylperoxide (BPO) used in conjunction with a tertiary amine [SAL 05]. Adhesives that rely solely on chemical initiating systems are seldom found on the market, the tendency being to offer dual activating mechanisms.

10.5. About the lifespan of dental bonding

While the clinical success of bonded dental restoration is currently regarded as satisfactory, it is known that degradation of the bonded joint occurs over time under the influence of physicochemical and biologic factors.

The former is particularly true for SES which, due to the presence of hydrophilic compounds in their very composition, are more sensitive to hydrolysis [VAN 11].

As for the latter, the heterogeneity and water content of dentin ECM that is infiltrated during the bonding process implies a weaker adhesion over time compared to enamel [DE 05, BRE 08]. Matrix metalloproteases (MMPs) are the major enzymes involved in the degradation of the dentin ECM. These are a family of about 30 secreted and membrane-bound zinc-endopeptidases functioning at neutral pH, requiring Ca^2+ for activity and collectively capable of degrading all the components of the ECM such as fibrillar and non-fibrillar collagens, fibronectin, laminin elastin and basement membrane glycoproteins [BIR 93, KOH 94, TAK 95, VIS 03]. Their structure, functions and biochemistry have been widely documented, while their role in the formation, maintenance and functioning of the dentin-pulp complex has also been investigated [VIS 03]. MMP-2, -8, -9 and -13 have also been identified in sound and carious dentin [TJA 98, MAR 00, SUL 04, CHA 06, MAZ 07, SUL 07].

Released and activated proteinases are thought to be responsible for the degradation of collagen fibrils in poorly infiltrated demineralized dentin *in vitro* [PAS 04, TAY 06, CAR 07a]. The degradation of collagen at the bottom of the hybrid layer has also been confirmed *in vivo* [KOS 04, HEB 05].

The MMP presence induced by the use of acidic adhesive systems, whether they were ER or SES has also been proven [MAZ 06, NIS 06]. Lehmann *et al.* showed in 2009, on tooth slice culture, that self-etching adhesives stimulate the secretion of MMP from the pulp-dentin complex by odontoblastes (dentin forming cells) implying that these participate in the degradation of the hybrid layer over time. Their results suggest that adhesive systems might benefit from the incorporation of effective, non-toxic inhibitors of MMPs or molecules blocking their expression by odontoblasts [LEH 09].

For that matter, chlorhexidine has been proposed as an add-on to the bonding protocol, for etch and rinse adhesive systems. Bond durability was enhanced by its use while immediate bonding performance was not affected [HEB 05, CAR 07a, CAR 07b]. Other synthetic inhibitors have been proposed like galardin or benzalkonium chloride with satisfactory results [BRE 10, TEZ 10]. However, modifying the application protocol of adhesive systems by adding another step and another product to use adds complexity, which goes against the current trend toward simpler and easier to use bonding systems.

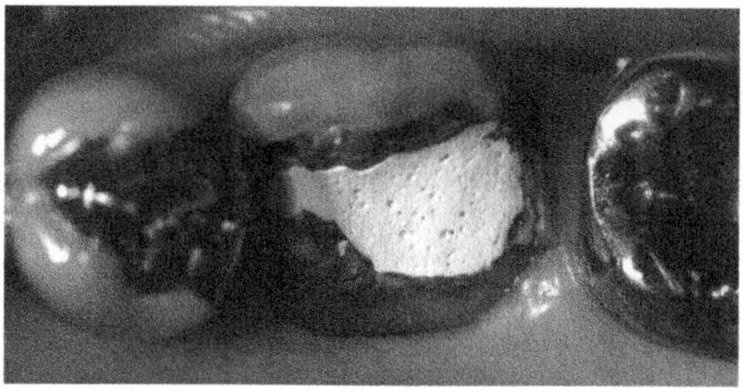

Figure 10.6. *Initial clinical view of the damaged teeth*

10.6. Clinical illustration

The use of adhesive techniques allows the practitioner to achieve durable and esthetic results while preserving sound tissues that would have been lost if a legacy method was used. We illustrate this statement with a clinical situation presented below.

The initial situation is illustrated with a huge loss of dental substance on a lower molar (Figure 10.6). The use of a classic restoration technique would imply a root canal treatment, followed by the placement of root anchoring then a total crown that would need the removal of a large amount of dental tissue. The radiogram, shown in Figure 10.7, allows us to assess the fact that a root canal is not needed as no sign of tooth necrosis is visible.

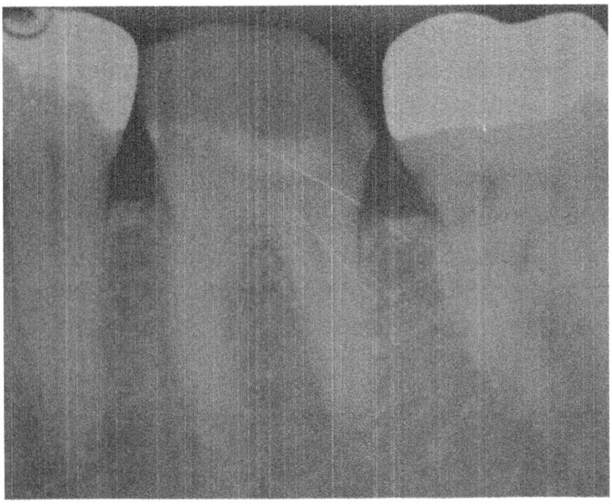

Figure 10.7. *Radiogram taken to assess the clinical situation and confirm that no sign of necrosis was present*

Current scientific data suggest that maximum tissue preservation is the best way to treat these damaged teeth. Decayed tissues are first removed, the composite is then bonded on the residual dentinal structure to ensure a good seating of the bonded prosthetic restoration

(Figure 10.8). An impression is made and transmitted to a technician who makes the prosthetic element shown in Figure 10.9.

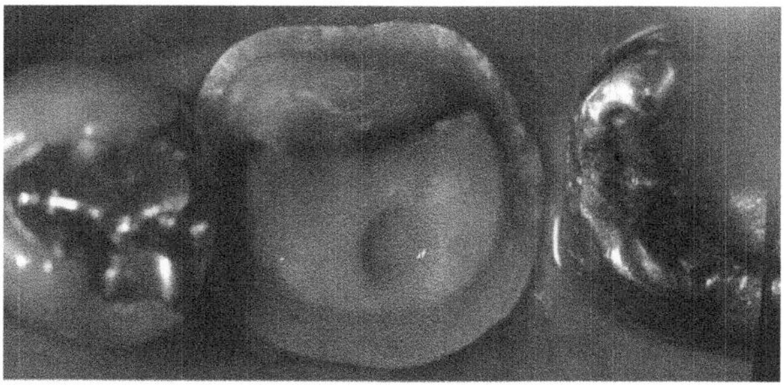

Figure 10.8. *View of the tooth after preparation. Composite was bonded on dentin and a very thin layer of enamel was removed (upper) to ensure good mechanical properties of the prosthetic restoration*

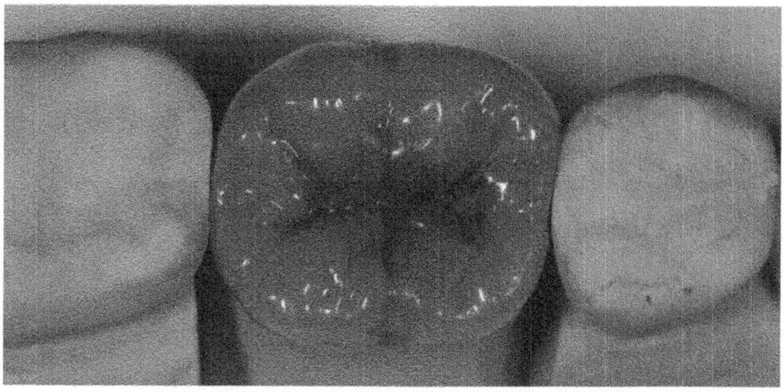

Figure 10.9. *Prosthetic element to be bonded seated on a cast mold of the dental arch*

The prosthetic element is seated on the tooth insulated by a rubber dam to ensure good hygiene and a contaminant-free environment while following the bonding procedure (Figure 10.10).

The final view of the restored teeth is presented in Figure 10.11. Through the use of adhesive dentistry, no sound tissues were removed aside from what was needed to ensure adequate seating of the restoration while tooth vitality was preserved, and function and esthetic were totally restored.

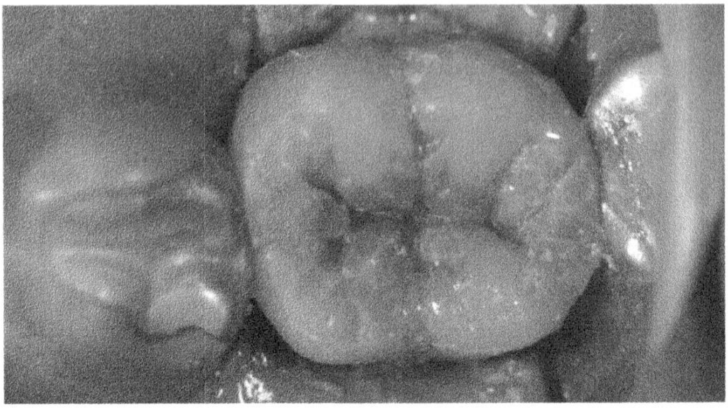

Figure 10.10. *Bonding of the prosthetic element. Tooth is insulated from oral environment with a rubber dam*

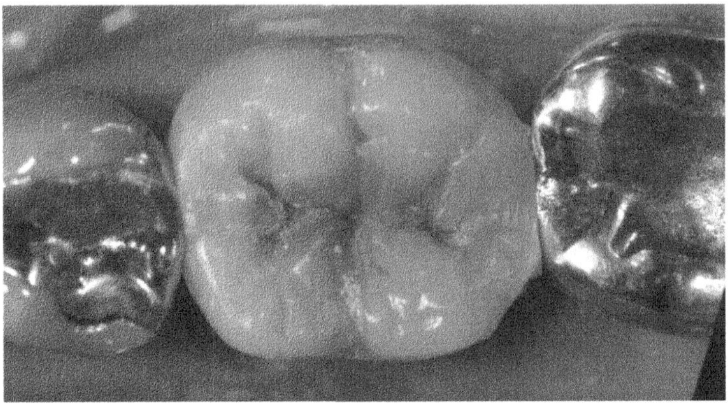

Figure 10.11. *Final view of the restored tooth*

10.7. Acknowledgments

The authors would like to dedicate this chapter to the loving memory of Professor Michel Degrange (1946–2010), our great teacher and mentor, the memory of whom we try to honor by continuing his work, pursuing his quest toward adhesive, minimally invasive, esthetic and ethical dentistry.

10.8. Bibliography

[AKI 07] AKIMOTO N., TAKAMIZU M., MOMOI Y., "10-year clinical evaluation of a self-etching adhesive system", *Operative Dentistry*, vol. 32, no. 1, pp. 3–10, 2007.

[BER 91] BERTOLOTTI R.L., "Total etch–the rational dentin bonding protocol", *Journal of Esthetic Dentistry*, vol. 3, no. 1, pp. 1–6, 1991.

[BIR 93] BIRKEDAL-HANSEN H., MOORE W.G., BODDEN M.K., *et al.*, "Matrix metalloproteinases: a review", *Critical Reviews in Oral Biology and Medicine*, vol. 4, no. 2, pp. 197–250, 1993.

[BRE 08] BRESCHI L., MAZZONI A., RUGGERI A., *et al.*, "Dental adhesion review: aging and stability of the bonded interface", *Dental Materials*, vol. 24, no. 1, pp. 90–101, 2008.

[BRE 10] BRESCHI L., MARTIN P., MAZZONI A., *et al.*, "Use of a specific MMP-inhibitor (galardin) for preservation of hybrid layer", *Dental Materials*, vol. 26, no. 6, pp. 571–578, 2010.

[BUO 55] BUONOCORE M.G., A simple method of increasing the adhesion of acrylic filling materials to enamel surfaces", *Journal of Dental Research*, vol. 34, no. 6, pp. 849–853, 1955.

[BUO 67] BUONOCORE M.G., "Principles of adhesive retention and adhesive restorative materials", *Journal of the American Dental Association*, vol. 67, no. 1963, pp. 382–391, 1967.

[CAR 07a] CARRILHO M.R., CARVALHO R.M., DE GOES M.F., *et al.*, "Chlorhexidine preserves dentin bond in vitro", *Journal of Dental Research*, vol. 86, no. 1, pp. 90–94, 2007.

[CAR 07b] CARRILHO M.R., GERALDELI S., TAY F., *et al.*, "In vivo preservation of the hybrid layer by chlorhexidine", *Journal of Dental Research*, vol. 86, no. 6, pp. 529–533, 2007.

[CHA 06] CHAUSSAIN-MILLER C., FIORETTI F., GOLDBERG M., *et al.*, "The role of matrix metalloproteinases (MMPs) in human caries", *Journal of Dental Research*, vol. 85, no. 1, pp. 22–32, 2006.

[DE 05] DE MUNCK J., VAN LANDUYT K., PEUMANS M., *et al.*, "A critical review of the durability of adhesion to tooth tissue: methods and results", *Journal of Dental Research*, vol. 84, no. 2, pp. 118–132, 2005.

[DEG 05] DEGRANGE M., "Les systèmes adhésifs amélo-dentinaires", *Realités Cliniques*, vol. 16, no. 4, pp. 327–348, 2005.

[FUS 79] FUSAYAMA T., NAKAMURA M., KUROSAKI N., *et al.*, "Non-pressure adhesion of a new adhesive restorative resin", *Journal of Dental Research*, vol. 58, no. 4, pp. 1364–1370, 1979.

[FUS 80] FUSAYAMA T., *New Concepts in Operative Dentistry*, Quintessence Publishing Co., Inc., Tokyo, 1980.

[HEB 05] HEBLING J., PASHLEY D.H., TJADERHANE L., *et al.*, "Chlorhexidine arrests subclinical degradation of dentin hybrid layers *in vivo*", *Journal of Dental Research*, vol. 84, no. 8, pp. 741–746, 2005.

[KOH 94] KOHN E.C., JACOBS W., KIM Y.S., *et al.*, "Calcium influx modulates expression of matrix metalloproteinase-2 (72-kDa type IV collagenase, gelatinase A)", *Journal of Biological Chemistry*, vol. 269, no. 34, pp. 21505–21511, 1994.

[KOS 04] KOSHIRO K., INOUE S., TANAKA T., *et al.*, "In vivo degradation of resin-dentin bonds produced by a self-etch vs. a total-etch adhesive system", *European Journal of Oral Science*, vol. 112, no. 4, pp. 368–375, 2004.

[LEH 09] LEHMANN N., DEBRET R., ROMÉAS A., *et al.*, "Self-etching increases matrix metalloproteinase expression in the dentin-pulp complex", *Journal of Dental Research*, vol. 88, pp. 77–82, 2009.

[MAR 00] MARTIN-DE LAS HERAS S., VALENZUELA A., OVERALL C.M., "The matrix metalloproteinase gelatinase A in human dentine", *Archives of Oral Biology*, vol. 45, no. 9, pp. 757–765, 2000.

[MAZ 06] MAZZONI A., PASHLEY D.H., NISHITANI Y., *et al.*, "Reactivation of inactivated endogenous proteolytic activities in phosphoric acid-etched dentine by etch-and-rinse adhesives", *Biomaterials*, vol. 27, no. 25, pp. 4470–4476, 2006.

[MAZ 07] MAZZONI A., MANNELLO F., TAY F.R., *et al.*, "Zymographic analysis and characterization of MMP-2 and -9 forms in human sound dentin", *Journal of Dental Research*, vol. 86, no. 5, pp. 436–440, 2007.

[NAG 06] NAGASE H., VISSE R., MURPHY G., "Structure and function of matrix metalloproteinases and TIMPs", *Cardiovascular Research*, vol. 69, no. 3, pp. 562–573, 2006.

[NAK 98] NAKABAYASHI N., PASHLEY D.H., *Hybridization of Dental Hard Tissues*, Quintessence, London, Tokyo, 1998.

[NIS 06] NISHITANI Y., YOSHIYAMA M., WADGAONKAR B., *et al.*, "Activation of gelatinolytic/collagenolytic activity in dentin by self-etching adhesives", *European Journal of Oral Science*, vol. 114, no. 2, pp. 160–166, 2006.

[PAR 99] PARK Y.J., CHAE K.H., RAWLS H.R., "Development of a new photoinitiaton system for dental light-cure composite resins", *Dental Materials*, vol. 15, no. 2, pp. 120–127, 1999.

[PAS 04] PASHLEY D.H., TAY F.R., YIU C., *et al.*, "Collagen degradation by host-derived enzymes during aging", *Journal of Dental Research*, vol. 83, no. 3, pp. 216–221, 2004.

[PAS 11] PASHLEY D.H., TAY F.R., BRESCHI L., *et al.*, "State of the art etch-and-rinse adhesives", *Dental Materials*, vol. 27, no. 1, pp. 1–16, 2011.

[PER 03] PERDIGAO J., GERALDELI S., HODGES J.S., "Total-etch versus self-etch adhesive: effect on postoperative sensitivity", *Journal of the American Dental Association*, vol. 134, no. 12, pp. 1621–1629, 2003.

[PEU 05] PEUMANS M., KANUMILLI P., DE MUNCK J., *et al.*, "Clinical effectiveness of contemporary adhesives: a systematic review of current clinical trials", *Dental Materials*, vol. 21, no. 9, pp. 864–881, 2005.

[PEU 10] PEUMANS M., DE MUNCK J., VAN LANDUYT K.L., *et al.*, "Eight-year clinical evaluation of a 2-step self-etch adhesive with and without selective enamel etching", *Dental Materials*, vol. 26, no. 12, pp. 1176–1184, 2010.

[SAL 05] SALZ U., ZIMMERMANN J., SALZER T., "Self-curing, self-etching adhesive cement systems", *Journal of Adhesive Dentistry*, vol. 7, no. 1, pp. 7–17, 2005.

[SUL 04] SULKALA M., PAAKKONEN V., LARMAS M., *et al.*, "Matrix metalloproteinase-13 (MMP-13, collagenase-3) is highly expressed in human tooth pulp", *Connective Tissue Research*, vol. 45, nos. 4–5, pp. 231–237, 2004.

[SUL 07] SULKALA M., TERVAHARTIALA T., SORSA T., *et al.*, "Matrix metalloproteinase-8 (MMP-8) is the major collagenase in human dentin", *Archive of Oral Biology*, vol. 52, no. 2, pp. 121–127, 2007.

[TAK 95] TAKINO T., SATO H., SHINAGAWA A., *et al.*, "Identification of the second membrane-type matrix metalloproteinase (MT-MMP-2) gene from a human placenta cDNA library. MT-MMPs form a unique membrane-type subclass in the MMP family", *Journal of Biological Chemistry*, vol. 270, no. 39, pp. 23013–23020, 1995.

[TAY 02] TAY F.R., KING N.M., CHAN K.M., *et al.*, "How can nanoleakage occur in self-etching adhesive systems that demineralize and infiltrate simultaneously?", *Journal of Adhesive Dentistry*, vol. 4, no. 4, pp. 255–269, 2002.

[TAY 06] TAY F.R., PASHLEY D.H., LOUSHINE R.J., *et al.*, "Self-etching adhesives increase collagenolytic activity in radicular dentin", *Journal of Endodonty*, vol. 32, no. 9, pp. 862–868, 2006.

[TEZ 10] TEZVERGIL-MUTLUAY A., MUTLUAY M.M., GU L.S., *et al.*, "The anti-MMP activity of benzalkonium chloride", *Journal of Dentistry*, vol. 39, no. 1, pp. 57–64, 2010.

[TJA 98] TJADERHANE L., LARJAVA H., SORSA T., *et al.*, "The activation and function of host matrix metalloproteinases in dentin matrix breakdown in caries lesions", *Journal of Dental Research*, vol. 77, no. 8, pp. 1622–1629, 1998.

[UNE 04] UNEMORI M., MATSUYA Y., AKASHI A., *et al.*, "Self-etching adhesives and postoperative sensitivity", *American Journal of Dentistry*, vol. 17, no. 3, pp. 191–195, 2004.

[VAN 01] VAN MEERBEEK B., VARGAS M., INOUE S., *et al.*, "Adhesives and cements to promote preservation dentistry", *Operative Dentistry*, vol. 6, pp. 119–144, 2001.

[VAN 03] VAN MEERBEEK B., DE MUNCK J., YOSHIDA Y., *et al.*, "Buonocore memorial lecture. Adhesion to enamel and dentin: current status and future challenges", *Operative Dentistry*, vol. 28, no. 3, pp. 215–235, 2003.

[VAN 07a] VAN DIJKEN J.W., SUNNEGARDH-GRONBERG K., LINDBERG A., "Clinical long-term retention of etch-and-rinse and self-etch adhesive systems in non-carious cervical lesions. A 13 years evaluation", *Dental Materials*, vol. 23, no. 9, pp. 1101–1107, 2007.

[VAN 07b] VAN LANDUYT K.L., SNAUWAERT J., DE MUNCK J., *et al.*, "Systematic review of the chemical composition of contemporary dental adhesives", *Biomaterials*, vol. 28, no. 26, pp. 3757–3785, 2007.

[VAN 11] VAN MEERBEEK B., YOSHIHARA K., YOSHIDA Y., *et al.*, "State of the art of self-etch adhesives", *Dental Materials*, vol. 27, no. 1, pp. 17–28, 2011.

[VIS 03] VISSE R., NAGASE H., "Matrix metalloproteinases and tissue inhibitors of metalloproteinases – structure, function, and biochemistry", *Circulation Research*, vol. 92, no. 8, pp. 827–839, 2003.

[YOS 00] YOSHIDA Y., VAN MEERBEEK B., NAKAYAMA Y., *et al.*, "Evidence of chemical bonding at biomaterial-hard tissue interfaces", *Journal of Dental Research*, vol. 79, no. 2, pp. 709–714, 2000.

[YOS 04] YOSHIDA Y., NAGAKANE K., FUKUDA R., *et al.*, "Comparative study on adhesive performance of functional monomers", *Journal of Dental Research*, vol. 83, no. 6, pp. 454–458, 2004.

11

Glass Ionomer Cements: Application in Pediatric Dentistry

11.1. Introduction

Glass ionomer cements (GICs) are polyalkenoates cements. The term "glass" refers to the nature of the particles used, based on fluoro-alumino-silicates, often as a milled fine powder. It can be decomposed by acid attack and can then release the ions forming the cement, typically calcium and aluminum ions and, optionally, strontium, lanthanide or zinc ions, according to the composition. The term "ionomer" refers to a polymer composed of macromolecular chains, with a small proportion of ionized or ionizable groups (usually 5–10%) cross-linked by ionic bridges. This is often polyacrylic acid, but it may also contain polymers and co-polymers of acrylic, itaconic, maleic or vinyl phosphoric acid. All GICs are the result of an acid-base reaction resulting from the mixing of the fluoro-alumino-silicate glass powder (base) and an aqueous solution of polyacrylic acid (acid), forming a polysalt which surrounds the glass particles which have not completely reacted (Figures 11.1 and 11.2).

This composition and structure give GICs unique properties such as a spontaneous adhesion to dental tissues, a dimensional stability to moisture, a coefficient of thermal expansion similar to the dental

---

Chapter written by Elisabeth DURSUN, Stéphane LE GOFF and Jean-Pierre ATTAL.

tissues, no shrinkage and a bio-activity by the anti-cariogenic action of fluoride release.

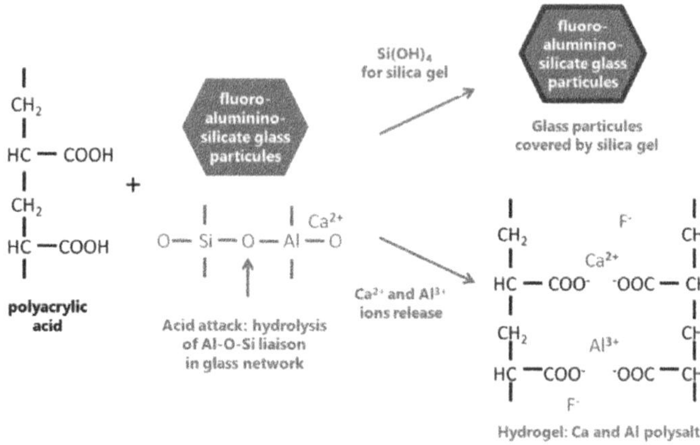

Figure 11.1. *The setting of GICs involves an acid-base reaction between polyalkenoic acid and fluoro-alumino-silicate glass particles*

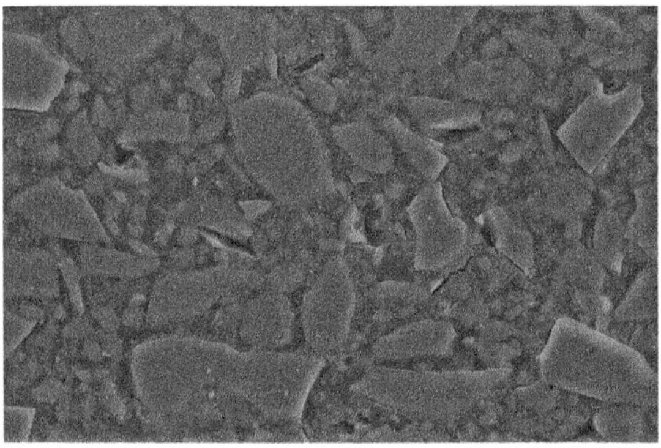

Figure 11.2. *The structure of glass ionomer cements – a representative SEM image*

11.2. Resin-modified and high viscous glass ionomer cements

In order to increase the mechanical properties of GIC, it was proposed to incorporate resin components (generally monomers of hydroxyethyl methacrylate (HEMA)). This new family of GIC was named resin-modified glass ionomers (RMGICs). From a chemical perspective, there are two ways to obtain an RMGIC:

– by adapting the matrix resin to the GIC matrix so that both lead to an interpenetrating network [ANT 87];

– by partially modifying the polyacid by grafting it, namely polymerizable groups (by esterification with HEMA), while the residual carboxylic groups facilitate the acid-base reaction with the glass [MIT 91].

The RMGICs activated by photo and chemical polymerization (Figure 11.3), in addition to the acid-base reaction, have been described as "dual-cure" or "tri-cure" to indicate a continuing polymerization under the light source.

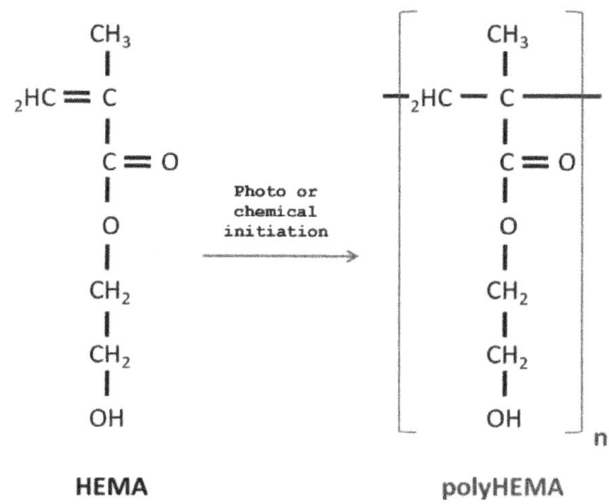

Figure 11.3. *The polymerization reaction*

High viscous (or condensable) glass ionomer cements (HVGICs) were also developed. These materials set faster and are of higher viscosity because of fine glass particles, anhydrous polyacrylic acids of high molecular weight and a high powder-to-liquid mixing ratio. The setting reaction is the same as the acid-base reaction of typical conventional GICs [FRA 97].

Some examples of brands of commercial materials are:

– RMGICs: Fuji II LC (GC), Riva Light Cure (SDI), Photac-Fil (3M-Espe) and Ionolux (Voco);

– HVGICs: Fuji IX (GC), Riva Self Cure (SDI), HiFi (Shofu), Ketac Molar (3M-ESPE), Chemfil Rock (Dentsply) and Ionofil Molar (Voco).

The main differences between these two types of material relate to their mechanical properties and implementation. HVGICs have the advantage of single-step placement (particularly attractive property for proximal cavities) and, in certain formulations, have accelerated chemical bonding. However, they are not robust in the medium term in proximal areas [QVI 10]. Limiting their use proximally for less than two to three years in the dental arch, and also for their use in small- to medium-sized cavities, is therefore recommended [FOR 03]. Nevertheless, it is possible that the use of a protective varnish (G-Coat Plus, GC) may considerably improve durability [FRI 11]. However, one might question how bioactive fluoride-releasing properties are maintained when a protective varnish is used.

Finally, it should be noted that a new high-viscosity RMGIC is now available (HV Riva Light Cure, SDI); this is an RMGIC that can be used as an HVGIC.

11.3. Dental adhesion and surface treatments

One of the most interesting properties of RMGICs is their ability to adhere naturally to dental tissues. Their bond is essentially chemical rather than micromechanical and involves ionic and hydrogen bonds. These chemical bonds form a tight and durable sealing, protecting and preventing the pulp risk of secondary caries.

Adhesion to collagen would essentially be by hydrogen and ionic bonds, with the structure of the apatite of the dentine. Adhesion to the mineral part would follow:

– a dynamic process of phosphate and calcium ion exchange between COO^- groups of the polyacrylate and the dentin of hydroxyapatite, with the formation of an intermediate layer of polyacrylates, particularly rich in calcium and phosphate ions;

– direct bonds to calcium ions of the apatite [WIL 88].

Although GICs are considered "self-adhesive" to dental tissues, the dental surface is not without effect on the adhesion performance. The dental tissues are always covered by the smear layer (drilling residue) that impedes contact with GIC. Hence, conditioning is required to remove it (Figures 11.4 and 11.5). A number of acids or products have been proposed to prepare the dental surface.

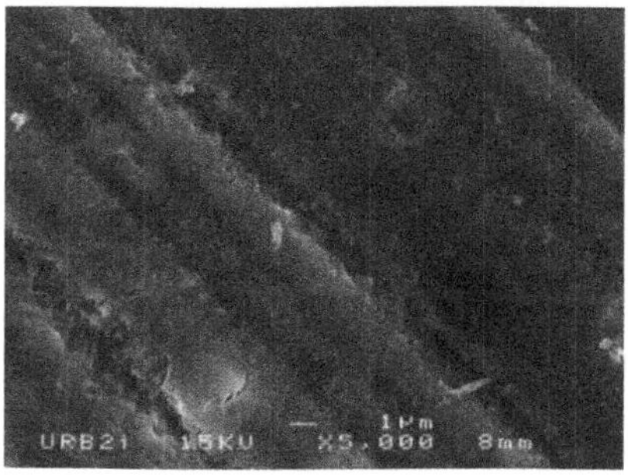

Figure 11.4. *Dentin surface without conditioning*

An application of polyacrylic acid (10–20%) seems to be the most reliable solution compared to phosphoric or citric acid (too demineralizing). An interesting alternative is the use of ethylenediaminetetraacetic acid (EDTA).

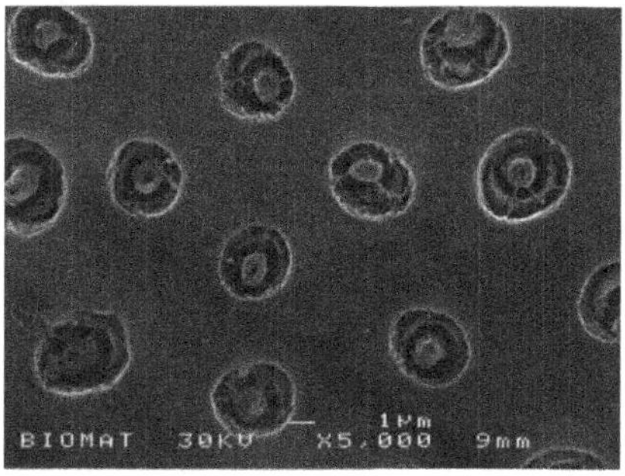

Figure 11.5. *Dentin after polyacrylic acid conditioning (20%)*

Otherwise, the use of a self-etching adhesive in combination with an RMGIC (Figure 11.6) increases the shear bond strength and this combination would be tolerant to moisture contamination [DUR 11].

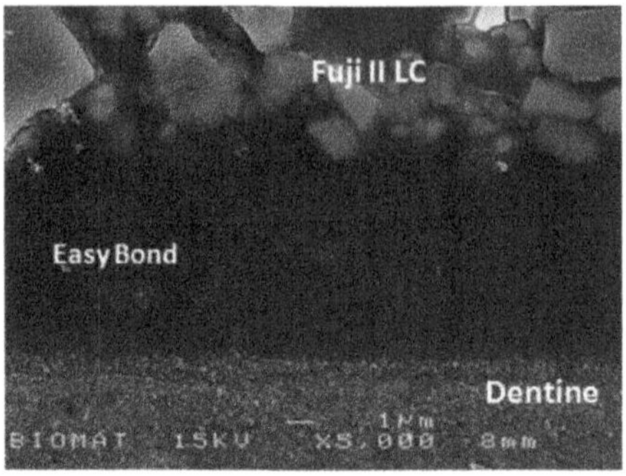

Figure 11.6. *A combination of a self-etching adhesive (Easy Bond, 3M ESPE) with an RMGIC (Fuji II, LC, GC) bonded to dentin – a representative SEM image*

11.4. Glass ionomers: application in pediatric dentistry

11.4.1. *Indications*

Successful restoration is linked to various factors: the material, the practitioner and the patient [DON 06]. The latter characterizes the specificity of pediatric dentistry. The limited cooperation of young patients justifies the use of materials that can be easily manipulated with simple protocol.

Furthermore, primary teeth are distinguished from permanent teeth mainly by their anatomy and their limited time in the dental arch. Consequently, even if the practitioner has the same array of materials for permanent teeth as for primary teeth (composite resins, amalgams, compomers and GICs), the specificities for the restoration of primary teeth are different.

Primary teeth are characterized by a thin layer of enamel consisting of enamel prisms that are directed vertically to the proximal surface. In the case of carious lesions, this tenuity can lead to extensive destruction, exacerbated by the fact that the prisms have poor cohesion.

Dentin forms an equally thin layer and its wide tubules facilitate bacterial penetration, accelerating the risk of pulp contamination. It is therefore important to work with sealable restorative materials.

The pulp chamber is proportionally much bigger than in permanent teeth and the pulpal horns are prominent. A carious lesion can therefore occur rapidly close to the pulp. It is therefore important to work with adhesive materials that do not require secondary cavity retention forms that may decay and cause pulpal exposure. For the same reason, smooth surfaces in the youngest patients which are affected by linear enamel caries or early carious lesions in the occlusal grooves or proximal surfaces of molar teeth [PSO 03, PSO 09] call for minimally invasive adhesive dentistry.

Owing to their short crown height, marked cervical constriction, relations with adjacent teeth and large gingival papillae, primary teeth can cause difficulties in establishing an isolated operative field, rendering the use of hydrophobic materials problematic [BUR 02]. It is therefore important to work with hydrophilic material.

Proximal caries adjacent to the primary tooth under treatment are common. Fluoride-releasing material placed on the proximal surface of the restoration could be advantageous in patients with a controlled risk of caries, to reduce the development and progression of caries on the proximal surface of the adjacent tooth. It is therefore important to work with bioactive material [QVI 10].

Moreover, sometimes the tooth's short remaining time in the arch may admit the use of materials compatible with this duration. Additionally, as masticatory constraints in children are lower than in adults [BRA 96, CAS 10, PAL 10], materials that are relatively less mechanically resistant may prove to be suitable. Thus, while materials with mechanical properties are crucial for permanent teeth, materials with lower mechanical properties may suffice for primary teeth in certain situations. This explains why GICs, markedly less mechanically resistant than composites, may have a role in pediatric dentistry.

Therefore, besides the need for fast implementation related to the patient's age, restorative material for primary teeth should also be sealable and adhesive to tooth tissues, bioactive and hydrophilic.

Glass ionomers meet all of these requirements.

11.4.2. *Longevity of restorative materials in primary teeth*

A review of the literature concerning the longevity of dental materials used in primary dentition highlights a wide variation in success rates. Indeed, numerous factors are involved: the type and brand of the material used, the practitioner's experience, the site and depth of the carious lesion and the age and co-operation of the patient.

Additionally, the lifespan of restorations in primary teeth is significantly different from that of permanent teeth, regardless of the chosen material [HIC 99]. This emphasizes the specificity of the selection criteria for primary dentition material.

Yengopal and colleagues in 2009 [YEN 09] conducted a systematic review of the literature, comparing the outcomes of different materials used for the restoration of primary teeth, in terms of pain relief, durability and esthetics. The study concluded that, from 1996 to 2009, there were only two well-conducted randomized clinical trials, evaluating the different restorative materials. These trials reported no significant differences between the materials. In one of these two trials, Donly and colleagues in 1999 [DON 99] compared an RMGIC with amalgam over a three-year period. However, owing to the high "lost to follow-up" rate, only the 12-month results are reported. No significant difference was found.

In terms of longevity, GICs are therefore materials that may pose an alternative to amalgams or composites for the restoration of primary teeth for a limited period of time. At present, two types of GIC are clinically relevant: RMGICs and HVGICs. However, some studies demonstrate differences in longevity depending on the type of GIC used and the site (occlusal or proximal) of the cavity.

11.4.3. *Examples of clinical cases*

Whatever the clinical situation, an operative field will always be established whenever possible. For the following clinical case, an isolated operative field was established. It should be noted that, with or without an operative field, the moisture tolerance of RMGICs, but also their bioactive nature and fluoride release, gives them an advantage over adhesive materials.

Clinical case: example of the restoration of a cervical lesion on a primary tooth with an RMGIC: Fuji II ® LC (GC). Figures 11.7, 11.8 and 11.9 present the restoration of an inferior canine with an RMGIC.

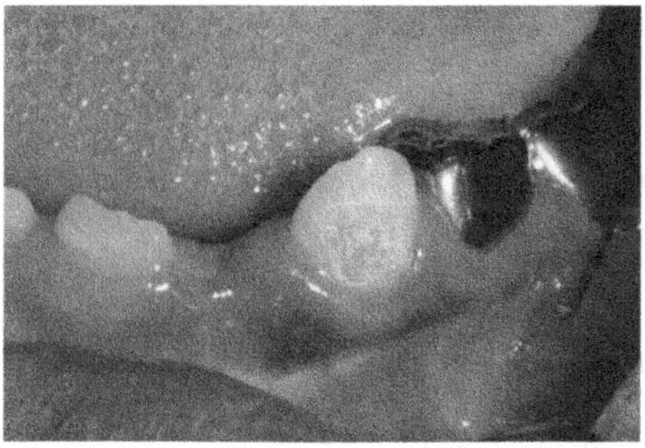

Figure 11.7. *Initial clinical view – decay on an inferior canine*

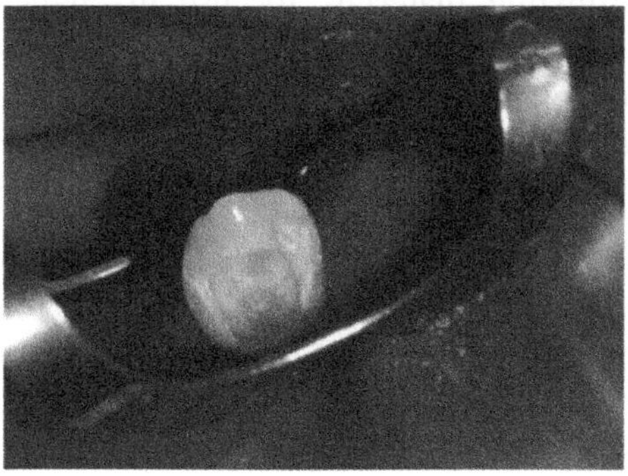

Figure 11.8. *Isolation of the tooth by means of an operative field and carie removal (then, polyacrylic acid was applied, rinsed and dried)*

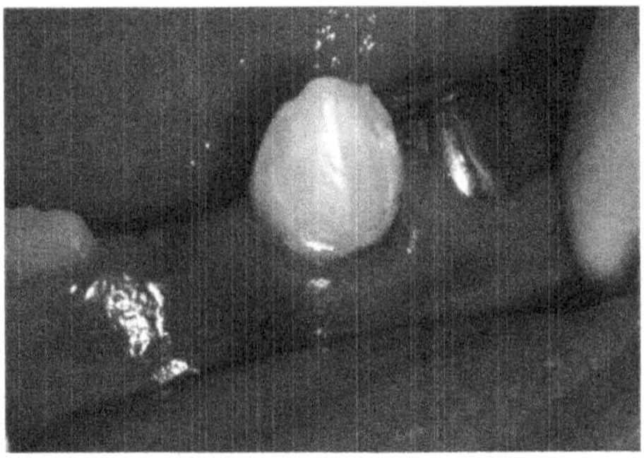

Figure 11.9. *Filling of the coronal cavity with Fuji II®*

In this case involving a juxta-gingival buccal lesion, RMGIC was the appropriate procedure. Admittedly, a composite restoration could have been carried out as the operative field could be established. However, the protocol for a composite is more time consuming and this material does not release fluoride.

11.5. Conclusion

The principal characteristics of glass ionomers include the ability to adhere naturally to enamel and dentin, the cariostatic effect of fluoride release and the moisture tolerance. They are therefore particularly worthwhile materials for use in challenging situations when isolation is impossible to obtain or for clinical situations concerning uncooperative children. In this regard, either an RMGIC or an HVGIC would be used when mechanical stress will be high.

11.6. Bibliography

[ANT 87] ANTONUCCI J.M., "Formulation and evaluation of resin modified glass ionomer cements", *Transactions of the 13th Annual Meeting*, Society of Biomaterials, New York, 1987.

[BRA 96] BRAUN S., HNAT W.P., FREUDENTHALER J.W., *et al.*, "A study of maximum bite force during growth and development", *The Angle Orthodontist*, vol. 66, no. 4, pp. 261–264, 1996.

[BUR 02] BURGESS J.O., WALKER R., DAVIDSON J.M., "Posterior resin-based composite: review of the literature", *Pediatric Dentistry*, vol. 24, no. 5, pp. 465–479, 2002.

[CAS 10] CASTELO P.M., PEREIRA L.J., BONJARDIM L.R., *et al.*, "Changes in bite force, masticatory muscle thickness, and facial morphology between primary and mixed dentition in preschool children with normal occlusion", *Annals of Anatomy*, vol. 23, no. 6, pp. 23–26, 2010.

[DON 06] DONOVAN T.E., "Longevity of the tooth/restoration complex: a review", *Journal of the California Dental Association*, vol. 34, no. 2, pp. 122–128, 2006.

[DON 99] DONLY K.J., SEGURA A., KANELLIS M., *et al.*, "Clinical performance and caries inhibition of resin-modified glass ionomer cement and amalgam restorations", *Journal of American Dental Association*, vol. 130, no. 10, pp. 1459–1466, 1999.

[DUR 11] DURSUN E., ATTAL J.P., "Combination of a self-etching adhesive and a resin-modified glass ionomer: effect of water and saliva contamination on bond strength to dentin", *Journal of Adhesive Dentistry*, vol. 13, no. 5, pp. 439–443, 2011.

[FOR 03] FORSS H., WIDSTRÖM E., "The post-amalgam era: a selection of materials and their longevity in the primary and young permanent dentitions", *International Journal of Paediatric Dentistry*, vol. 13, no. 3, pp. 158–164, 2003.

[FRA 97] FRANKENBERGER R., SINDEL J., KRAMER N., "Viscous glass-ionomer cements: a new alternative to amalgam in the primary dentition", *Quintessence International*, vol. 28, no. 10, pp. 667–676, 1997.

[FRI 11] FRIEDL K., HILLER K.A., FRIEDL K.H., "Clinical performance of a new glass ionomer-based restoration system: a retrospective cohort study", *Dental Materials*, vol. 27, no. 10, pp. 1031–1037, 2011.

[HIC 99] HICKEL R., MANHART J., *Advances in Glass Ionomer Cements, Glass-Ionomers and Compomers in Pediatric Dentistry*, Quintessence, Berlin, 1999.

[MIT 91] MITRA S.B., "Adhesion to dentin and physical properties of a light-cured glass-ionomer liner/base", *Journal of Dental Research*, vol. 70, no. 1, p. 72, 1991.

[PAL 10] PALINKAS M., NASSAR M.S., CECÍLIO F.A., *et al.*, "Age and gender influence on maximal bite force and masticatory muscles thickness", *Archives of Oral Biology*, vol. 55, no. 10, pp. 797–802, 2010.

[PSO 03] PSOTER W.J., ZHANG H., PENDRYS D.G., *et al.*, Classification of dental caries patterns in the primary dentition: a multidimensional scaling analysis", *Community Dentistry and Oral Epidemiology*, vol. 31, no. 3, pp. 231–238, 2003.

[PSO 09] PSOTER W.J., PENDRYS D.G., MORSE D.E., *et al.*, "Caries patterns in the primary dentition: cluster analysis of a sample of 5169 Arizona children 5-59 months of age", *International Journal of Oral Sciences*, vol. 1, no. 4, pp. 189–195, 2009.

[QVI 10] QVIST V., POULSEN A., TEGLERS P.T., *et al.*, "Fluorides leaching from restorative materials and the effect on adjacent teeth", *International Dental Journal*, vol. 60, no. 3, pp. 156–160, 2010.

[WIL 88] WILSON A.D., MCLEAN J.W., *Composition: Glass Ionomer Cement*, Quintessence Publishing, Chicago, pp. 21–42, 1988.

[YEN 09] YENGOPAL V., HARNEKER S.Y., PATEL N., *et al.*, "Dental fillings for the treatment of caries in the primary dentition", *The Cochrane Database of Systematic Reviews*, vol. 15, no. 2, pp. 1–32, 2009.

List of Authors

Jean ALIGN
Unité de Recherche en
Biomatériaux Innovants et
Interfaces (URB2i)
Paris Descartes University
Montrouge
France

Jean-Pierre ATTAL
Unité de Recherche en
Biomatériaux Innovants et
Interfaces (URB2i)
Paris Descartes University
Montrouge
France

Adeline DECAMBRON
Laboratoire de Bioingénierie et
Bioimagerie Ostéo-Articulaires
Paris Diderot University
and
Paris-Est University
Ecole Nationale Vétérinaire
d'Alfort
France

Mathieu DERBANNE
Unité de Recherche en
Biomatériaux Innovants et
Interfaces (URB2i)
Paris Descartes University
Montrouge
France

Elisabeth DURSUN
Unité de Recherche en
Biomatériaux, Innovation et
Interfaces (URB2i)
Paris Descartes University
Montrouge
France

Céline FALENTIN-DAUDRE
LBPS/CSPBAT
Institut Galilée
University of Paris 13
Villetaneuse
France

Stéphane Le GOFF
Unité de Recherche en
Biomatériaux Innovants et
Interfaces (URB2i)
Paris Descartes University
Montrouge
France

Amélie MAINJOT
Dental Biomaterials Research Unit
Department of Fixed
Prosthodontics
Institute of Dentistry
University of Liege (ULg)
Belgium
and
Biomaterials and Interfaces
Research Unit
Paris Descartes University
France

Mathieu MANASSERO
Laboratoire de Bioingénierie et
Bioimagerie Ostéo-Articulaires
Paris Diderot University
and
Paris-Est University
Ecole Nationale Vétérinaire
d'Alfort
France

Véronique MIGONNEY
Laboratoire de "Chimie,
Structures et Propriétés de
Biomatériaux et d'Agents
Thérapeutiques"
University of Paris 13
Villetaneuse
France

Géraldine ROHMAN
Laboratoire de Chimie,
Structures, Propriétés de
Biomatériaux et d'Agents
Thérapeutiques (CSPBAT)
University of Paris 13
Villetaneuse
France

Cédryck VAQUETTE
Institute of Health and
Biomedical Innovation
Queensland University of
Technology
Australia

Véronique VIATEAU
Laboratoire de Bioingénierie et
Bioimagerie Ostéo-Articulaires
Paris Diderot University
and
Paris-Est University
Ecole Nationale Vétérinaire
d'Alfort
France

Index

A, B

C, D

E, F, G, H

I, L, M

N, O, P

R, S

T, Z

Other titles from

ISTE

in

Bioengineering and Health Science

2014

DAO Tien Tuan, HO BA THO Marie-Christine
Biomechanics of the Musculoskeletal System: Modeling of Data Uncertainty and Knowledge

FARIÑAS DEL CERRO Luis, INOUE Katsumi
Logical Modeling of Biological Systems

TEBBANI Sihem, FILALI Rayen, LOPES Filipa, DUMUR Didier, PAREAU Dominique
CO2 Biofixation by Microalgae : Modeling, Estimation and Control

Printed and bound by CPI Group (UK) Ltd, Croydon, CR0 4YY
13/09/2026
14961076-0002